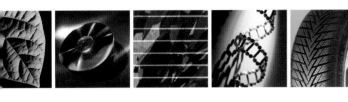

H.-J. Quadbeck-Seeger

**World of the Elements
Elements of the World**

Translated by José Oliveira

Author:
Prof. Dr. Hans-Jürgen Quadbeck-Seeger
67098 Bad Dürkheim, Germany

Artwork & design:
Gunther Schulz
67136 Fußgönheim, Germany

1. Edition 2007

Impressum:
All books published by Wiley-VCH are carefully produced. Nevertheless, authors, editors, and publisher do not warrant the information contained in these books, including this book, to be free of errors. Readers are advised to keep in mind that statements, data, illustrations, procedural details or other items may inadvertently be inaccurate.

Library of Congress Card No.:
applied for

British Library Cataloguing-in-Publication Data
A catalogue record for this book is available from the British Library.

Bibliographic information published by the Deutsche Nationalbibliothek
Die Deutsche Nationalbibliothek lists this publication in the Deutsche Nationalbibliografie; detailed bibliographic data are available in the Internet at <http://dnb.d-nb.de>.

© 2007 WILEY-VCH Verlag GmbH & Co. KGaA, Weinheim

All rights reserved (including those of translation into other languages). No part of this book may be reproduced in any form – by photoprinting, microfilm, or any other means – or transmitted or translated into a machine language without written permission from the publishers. Registered names, trademarks, etc. used in this book, even when not specifically marked as such, are not to be considered unprotected by law.

Typesetting: Gunther Schulz, Fußgönheim
Printing: betz-druck GmbH, Darmstadt
Binding: Litges & Dopf GmbH, Heppenheim

Printed in the Federal Republic of Germany
Printed on acid-free paper

ISBN: 978-3-527-32065-3

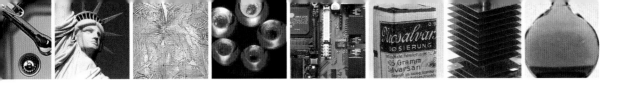

Hans-Jürgen Quadbeck-Seeger

World of the Elements
Elements of the World

With kind support from

◻ ▪ BASF

The Chemical Company

WILEY-VCH Verlag GmbH & Co. KGaA

Preface

Attention — this is not a textbook! It is also not meant to replace one. Nevertheless, there is a lot to be learnt, albeit in a different manner from that in which chemistry is normally presented. The initial question is old and simple: "What does the world consist of?" This leads to the next question: "Who were the researchers who discovered this?" They should not be forgotten by us, who take almost for granted all the advantages of progress. After all, the discovery of the 92 elements that occur in the universe and that can also be found on Earth is one of the greatest accomplishments of human intellectual curiosity. Through these discoveries we know what stars are made of, we know the composition of the Earth, and we know which elements are essential for life.

The transition of empirical alchemy in 18th century Europe to scientific chemistry allowed the discovery of more and more new elements through the thirst for knowledge, intuition, patience, and even luck. Known materials such as gold, silver, copper, iron, and lead were "suspected" to be elements relatively early. Despite all the best efforts, these materials could not be broken down into further components, and hence their being elements was consistent with the then generally recognized definition of John Dalton, which was also staunchly supported by Antoine de Lavoisier.

New scientific methods (e.g. electrolysis) allowed the veteran elements to be joined stepwise by more and more unknown and unexpected substances that fulfilled the criteria for an element. In 1869, after many attempts to bring order into the growing chaos, Dimitri Mendeleev revealed a daring concept with his Periodic Table and its predictions. Each of the then known elements was assigned a place. The gaps represented elements that were not yet known. The discoveries of such elements proved that there was an order and system to the elements. This order explained much that was previously puzzling, for instance, the different atomic radii observed that same year by Julius Lothar Meyer, which seemed to follow a periodic trend.

The representation of the periodic system in this book shows yet another perspective. Each element has not only its own history but also its own identity. This is determined by the number of protons in the nucleus (the atomic number) and the corresponding number of electrons in the atomic shell. These electrons, in turn, give each element their properties, their "personalities", so to speak. There are relationships, but each element is unique in the sum of its properties. The text describes the particularities of each element, and the chosen picture indicates a scene from everyday life where we would encounter the element. They are often hidden in functional systems such as electronics, or they impart particular properties to alloys, such as hardness and magnetism.

The tables and graphics at the end of the book provide an overview of how everything is connected. In general, it can be quite cumbersome for some to put together such a wealth of information. Hopefully, the selection presented will facilitate the search. Like all historical precedents, the discovery of the elements was complex and often multitracked. Some discoveries "were floating in the air" and were made independently by several researchers. Hence, even the authoritative literature leaves some questions unanswered, for example, regarding absolute priorities. These have been selected to the best of knowledge and belief, but unavoidable subjectivity should be borne in mind.

Chemistry would not be done justice if only the past and the status quo were discussed. Today, new heavy elements are discovered in nuclear accelerators as a result of their decomposition traces and are of interest in nuclear physics. The Periodic Table provides building blocks for new areas of chemistry. The possibilities for combining elements into defined compounds is far from exhausted, even though about 30 million have been described to date. Besides the question as to how molecules **react** with each other, a new phenomenon is becoming increasingly important: how molecules **interact** with each other. The principle of self-assembly was a condition for the origin of life. The targeted use of this fascinating property of materials to build functional systems is still in its infancy. An exciting phenomenon is the erratic change in properties as the particles of the material become very small. The door to the nanoworld (nano = 10^{-9}m) has only just been opened, but already fascinating perspectives with great potential can be seen in the transition from classical physics to quantum mechanics. Catalysis and materials science are undergoing dynamic developments. And finally, we are all aware of how molecular biology is rapidly developing at the interface of chemistry and biology. The Periodic Table is the foundation for the tower of knowledge and application of chemistry. The challenge for chemistry is and will remain the exploration of this knowledge for the good of mankind.

I am deeply grateful to many friends and colleagues, especially at BASF AG, for their help in collecting materials for this book. I am equally indebted to the imaginative graphic designer Gunther Schulz and the ever helpful colleagues at Wiley-VCH, without whose help this book would not have come together.

Hans-Jürgen Quadbeck-Seeger   *June 2007*

Table of Contents

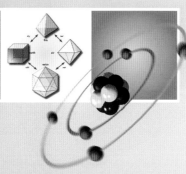
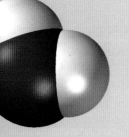

About the Historical Periodic Table and the Chemometer Data and Facts	6
The History of the Atomic Model	16
The Elements of the World Properties, Discovery, Occurance & Applications	25
Where to from Here?	83
The Elements and Life	84
The Elements in Culture and Language	88
Chemical Olympics Interesting Facts from the World of Chemistry	90
Fall of a Winged Word	95
Water	96

Appendix

Chronology of the Discovery of the Elements	98
Places of Discovery	99
Periodicity of Atomic Properties	100
Abundance of the Elements	102
World Production of Elements and Raw Materials	104
Nobel Prize Winners	106
Alternative Representations of the Periodic System	107
Occupation of the Shells and Orbitals	108
The World of the Elements — Literature	109
Sources of Photograph	110

About the **Historical Periodic Table** and the **Chemometer**

These two new representations arouse curiosity about the elements, their discovery, and their characteristics. And where can they be found? The close relationships between their properties and their importance in our lives and for civilization are made apparent.

Chemometer:
www.corporate.basf.com

▲ One of the first sketches by Dmitri Mendeleev regarding a new ordering of the elements.

About the Historical Periodic Table and the Chemometer: Data and Facts

The Linear Periodic System (Chemometer)

The linearization of the periodic system is actually a step back, as a decisive characteristic is thereby lost: the periodicity. And this is actually the key to understanding. But more about this soon. Nevertheless, a linear representation of the elements does make sense. Do I know all the names hidden behind the internationally accepted abbreviations? Where is a particular element that interests me? Which are its neighbors? These and other such questions can be quickly answered by consulting the chemometer.

The linear representation of the sequence of the elements has an interesting history. Upon the initiative of F. A. Kekulé, the first International Chemistry Congress took place in Karlsruhe in 1860 with 140 participants. An important outcome was the agreement on the atomic weights of the approximately 60 elements known at the time. However, the obvious ordering according to atomic weight seemed to make no "chemical" sense. Neighboring elements differed greatly, whereas some further apart behaved similarly. In 1864, the English chemist John Alexander Reina Newlands, who headed a laboratory in a sugar factory, noticed a remarkable similarity in the properties of elements eight places apart. His law of octaves at first only drew ridicule. Only after the physicochemical analysis of Julius Lothar Meyer (on the basis of the size of the atoms) and the visionary concept of Dimitri Mendeleev (according to the similarities in their chemical properties) was it recognized that Newlands was on the right track.

Towards the end of the 19th century, the ordering of the elements according to their atomic weight became of surprising importance. After Konrad Röntgen discovered X-rays in 1895, the talented young physicist Henry Moseley, working in the laboratories of Ernest Rutherford, investigated the X-ray spectra of all the then-known elements. His discovery was astonishing. The root of the frequency of the X-rays of each element was proportional to its atomic number (Moseley's Law). This was of groundbreaking significance for three reasons. First, Moseley ordered the elements in a sequence that confirmed the assumed order but also suggested gaps. This brought attention to a missing element (rhenium). Second, it showed that the phenomenon of periodicity, assigned to the entire kingdom of the elements by Mendeleev in a stroke of genius, covered both chemical as well as physical properties, such as atomic and ionic radii. Third, it showed once and for all that there were no further elements between atomic numbers 1 and 92; a few gaps remained to be filled, but otherwise the search could be called off. The world of the elements had been chartered, and each element was assigned a fixed position. Like many other talented youngsters, the promising Henry Moseley lost his life in World War I.

◀ *Henry Moseley (1887–1915) in the laboratory in which he discovered the law named after him.*

The Classical Periodic System

The discovery of the 92 natural elements, which occur in the universe and can all be found on Earth — not necessarily predictable — is one of the greatest achievements of the human intellectual curiosity. It seems a miracle: evolution gave rise to a being that was in a position to recognize the constitution of the world. And towards the end of the 20th century, this being even learnt to understand the principles of the processes of life to which it owes its existence. But that is a completely new chapter in the history of science. For now, let us stick to the elements.

Prehistoric metallurgy developed empirically. It was handed down from teacher to student and hence improved only slowly. In antiquity, seven elements (according to today's definition) were known: gold, silver, copper, tin, mercury, lead, and sulfur. Charcoal was not considered an element, and diamond was regarded as a mineral. Alchemy in ancient Arabia and in the Middle Ages in Europe was looked upon as a secretive science and set itself illusionary aims such as the conversion of lead into gold, the search for the philosopher's stone, or the quest for "alcahest", a substance that would dissolve anything. As no religion prescribed how or how many elements the respective Creator created, an abundance of myths and fantasies unfolded. It was Robert Boyle who in the spirit of the European Enlightenment brought about a paradigm shift in the 17th century. In his work "New Experiments", published in 1660, he described his experiments so clearly and in such detail that anyone could have reproduced them. This was, of course, his intention. He thus freed alchemy from the notorious veil of secrecy that brought it into such disrepute. For him it was no longer a question of personal gain in knowledge, but rather a general improved understanding. This marked the beginning of scientific chemistry in the open, free-thinking atmosphere of the Enlightenment. It is not coincidental that, without exception, all natural, stable elements were discovered or identified as such in Western Europe.

▲ *Robert Boyle (1627–1691).*

The early researchers were loners, sometimes even eccentric, such as the almost autistic, unsociable Henry Cavendish, then one of the richest men in England. They were driven by curiosity and a thirst for knowledge. They carried out their investigations independently and also financed them themselves. Those who were not wealthy worked under primitive conditions, for instance the idealistic provincial pharmacist Carl Wilhelm Scheele, who worked with no expensive apparatus, only common sense and memory, imagination and determination, talented hands, and at the cost of his health. Others put all their money into optimizing their equipment. In this way, the French researcher Antoine de Lavoisier developed chemistry into an exact science. The money he earned as a strict tax collector was used to acquire the best scales and measuring devices. He thereby justified the importance of precise methods to obtain dependable and reproducible statements on quantities. His life came to an early, tragic end: the French Revolution claimed his head in return for his dubious services to the king.

Early researchers had to contend with the disinterest and apathy of their environment. But their success awakened increasing interest. As chemistry started to show some application, it came to the attention of the powerful and the curious. The first laboratories with appointed positions in academies and universities came into

▲ Antoine de Lavoisier with his wife.

being at the beginning of the 19th century (starting in 1807 with the Royal Institution in London). Talented individuals from the less-wealthy classes joined the small community of researchers. Excellent examples include Humphry Davy, Michael Faraday, Martin Heinrich Klaproth, and Joseph Fraunhofer.

Subsequently, the knowledge of chemistry began to grow rapidly. The term "element" was defined by the humble Quaker and schoolmaster, John Dalton. Jöns Berzelius introduced the symbols for the elements. The discovery of the elements clearly occurred in phases. Usually the development of a method led to a new result. At first, the search was empirical and involved much guess work. The art of the blowpipe was mastered above all by the Swedish chemists. Antoine de Lavoisier then introduced exact measuring and, more importantly, balancing. Humphry Davy used the voltaic battery to do chemistry with electricity. By means of electrolysis he discovered five new elements. Robert Bunsen and Gustav Kirchhoff conducted their search with spectral analysis. The liquefaction of air by Carl Linde allowed Sir William Ramsey to discover all the noble gases, including radon. Radioactivity revealed the existence of unstable elements. And finally, Henry Moseley used X-ray spectra to confirm unambiguously the order of the elements.

This summary, however, plays down the difficulties encountered along the way. Chemists were confronted with a growing problem in the middle of the 19th century: the knowledge of chemistry was spreading like wildfire. Today, the situation would have been described as "complete chaos". Each element exhibited unique properties, which differed vastly from those of the neighboring elements, although their atomic weights were so close. One just needs to think of the three neighboring elements carbon, nitrogen, and oxygen.

Certainly, some similarities in the chemical properties of distant elements were recognized early on. J. W. Döbereiner was the first to draw attention to so-called "triads" (e.g. Ca, Sr, Ba). Some elements formed strong bases, others strong acids and these seemed to have an affinity for each other.

Then there were metals, nonmetals, and strange elements that behaved like metals. But where was the order? Had Isaac Newton not impres-

◀ The blowpipe is a metallic tube with a small nozzle and mouthpiece used to investigate minerals. Air is blown through the nozzle into the flame of a gas burner, causing the mineral sample to be heated to very high temperatures.

43	44	45	46	47	48	49	50	51	52	53	54
Tc	Ru	Rh	Pd	Ag	Cd	In	Sn	Sb	Te	I	Xe
98.91	101.1	102.9	106.4	107.9	112.4	114.8	118.7	121.8	127.6	126.9	131.3

▲ *"The creation" by William Blake (1794).*

sively shown 200 years before in physics that God wrote the book of Nature in the language of mathematics? Isaac Newton dedicated the later part of his life mainly to alchemy in the hope of also finding some order there. However, he failed miserably because the time was not ripe. The success of scientific research and the application of technology strengthened the belief that God created the world on the basis of mathematical laws. The famous 1974 painting by William Blake of the creation is a metaphor for the spirit of the time.

But what about the elements? How many were there? Could the elements then known to man not be further expanded. Could they only be found on Earth? Did God create them in the beginning or were they the result of some process of evolution? Even the existence of atoms was doubted. The nebulous concepts and the countless dead-ends led to the eminent French chemist Jean B. A. Dumas writing in 1837 to state that "If I had the power, I would strike the word "atom" from science".

There was a determined search for a system. Besides the suggestion of Newlands already described, almost 100 other attempts were registered. By a fortunate coincidence, we came into possession of a document that impressively captured the state of affairs of chemistry at the time. Several pictures were produced for the chapter "Chemical Technology" for the 11th edition of the Real-Enzyklopädie (1864–68) printed in 15 volumes by the publishers Brockhaus in Leipzig. One such picture depicted the lecture theater of the Chemical Institute of the University of Leipzig. Wilhelm Ostwald was later to teach in this room, and Carl Bosch as well as Alwin Mittasch graced this hall as students. Together, they later pioneered the industrial synthesis of ammonia at BASF. Let us concentrate on the Table of the Elements that adorns the front wall of the lecture hall. (Even today, no classroom in which chemistry is taught should be without a Periodic Table.)

The picture was drawn in 1864. The first International Chemistry Congress had taken place in Karlsruhe in 1860. The atomic weights had been agreed upon. Indium was discovered in 1863 and already added to the ranks of the 64 known elements. The discoverers still had the right to name the compounds they uncovered. And how were the elements ordered? The elements were listed ac-

◀ Lecture theater at the Chemical Institute of the University of Leipzig.

cording to the alphabet, which itself was created historically without any logic or connectivity. And yet there was a strong belief that some intrinsic order lay hidden in the world of the elements. Boldly inscribed above the Table was a biblical quotation from the Apocrypha of Solomon:

GOTT HAT ALLES NACH ZAHL MAASS UND GEWICHT GEORDNET (God ordered all things by measure and number and weight.)

The designers of the lecture room were, of course, proved correct. Only a few years later a systematic order was, indeed, recognized. An extraordinary double discovery was made in 1869. The German chemist Julius Lothar Meyer (1830–1895) noticed a remarkable periodicity during his rigorous scientific analysis of the atomic weights and volumes. He remained content with only a mild curiosity in this realization, as his interests lay primarily in physicochemical problems. He was objective and driven only by facts; he was wary of hypotheses

66	67	68	69	70	71	72	73	74	75	76	77
Dy	Ho	Er	Tm	Yb	Lu	Hf	Ta	W	Re	Os	Ir
162.5	164.9	167.3	168.9	173.0	175.0	178.5	180.9	183.8	186.2	190.2	192.2

▲ *Dmitri Ivanovich Mendeleev (1834–1907).*

and strictly rejected speculation. At the same time, a chemist from Saint Petersburg who was only known among people in the field, thought completely differently. Dimitri Mendeleev was blessed with a memory for chemical details. He was thus able to relate and connect chemical similarities between certain elements. His conclusions were indeed very bold. The ordering principle in the kingdom of the elements is the "periodicity". Who was this man, who looked so earthy and pensive. He hailed from a small village in Siberia. His mother recognized his extraordinary talents at a very young age and, in the face of many hardships, took the young boy to Saint Petersburg (she died soon thereafter) where he would have a chance of finding appropriate opportunities for his education. The versatile, gifted son dedicated himself to science, in particular chemistry. His extensive knowledge allowed him to match chemical properties and physical data, such as atomic weights and estimated atomic sizes, as if building a puzzle.

His periodic system did not meet with universal approval. This comes as no great surprise; today, such revolutionary ideas would be termed a "paradigm shift". Since the time of Isaac Newton and Gottfried Wilhelm von Leibniz, scientists had been to formulating scientific laws in equations. After all, had James Clerk Maxwell in a stroke of genius not very convincingly demonstrated the relationship between electricity and magnetism with an equation in 1855? In contrast, Mendeleev's system was mathematically very primitive. It involved a simple matrix in which the elements were distributed horizontally into periods and vertically into groups. Were the relationships really so simple?

It was the astounding coherence of the facts rather than intellectual persuasiveness that made the small chemistry community take the proposal seriously. But his system had several gaps, particularly since only 64 elements were known at the time. With Siberian stubbornness, Mendeleev predicted in 1871 that these gaps represented elements that were yet to be discovered. For example, the elements below aluminum and silicon were still missing. These he called "eka-aluminum" and "eka-silicon" and made predictions regarding their properties. When gallium was first found by Paul Emile Lecoq de Boibaudran in 1875 and germanium was discovered in 1886 by Clemens Winkler on the basis of these suspicions, there was no longer any doubt about the genius of Mendeleev's concept. Although the Nobel Prize was awarded from 1901 and Mendeleev lived until 1907, he was overlooked for this prize. "His" periodic system had in the meantime become common and obvious knowledge.

Chemists were not able to use their methods to determine the structure of the atom. The discovery of radioactivity by Henri Becquerel and the work of Marie and Pierre Curie showed, however, that heavy elements were not stable. The earlier postulate of their indivisibility could no longer be maintained. In 1906 Ernest Rutherford made the next horrorific revelation: his scattering experiments showed that the atom was almost empty. A tiny nuclear mass was circled by electrons at a large distance. For comparison, if the nucleus were the size of a cherry pit and were placed in the center of a football field, the electrons would be circulating in the back rows of the stadium. If the nucleus were the size of a football, the first electrons would be circling it at a distance of one kilometer. Between them would be absolute emptiness.

The physicists explained the phenomenon of uneven atomic weights as a consequence of the presence of varying numbers of neutral particles, neutrons, in the nucleus. Atoms that only differ in the number of neutrons are called isotopes and exhibit identical chemical behavior. The best-known isotopes, which also have their own names, are those of hydrogen. The deuterium nucleus consists of a proton and a neutron. Tritium atoms contain two neutrons and are unstable, that is, they are radioactive and decompose. Tritium has a half-life of 12.3 years. Stable and unstable isotopes of most elements are known. Seven elements of the Periodic Table are naturally radioactive, that is, none of their isotopes is stable. In nuclear fusion, a small amount of mass is converted into energy, so that a mass defect is observed. The international standard is the carbon atom with six protons, and six neutrons, which was defined as having an atomic weight of 12.000. As a consequence of the mass defect, the atomic weight of hydrogen is not exactly 1.000, but rather 1.008.

The lightning-quick, circulating electrons initially posed a serious problem to physicists. According to classical physics, moving charges generate an electromagnetic field; consequently they should lose energy and soon crash into the nucleus. The young Danish physicist Niels Bohr dedicated himself to this problem and postulated that if the electrons circulated in their appropriate orbitals, they would not obey the laws of classical physics. If the electrons received an appropriate amount of energy, they would jump to the next magical orbital. This planetary model was not only popular, but also explained the quantum jumps observed by Max Planck. It also fitted the theory of Albert Einstein published in 1905 that

▲ Henri Becquerel (1852–1908).

◀ Werner Heisenberg (1901–1976).

suggested that space and time form a continuum, whereas matter and energy were quantized, that is, "grainy". All was right with the world of chemistry for 20 years, until in 1927 the young Werner Heisenberg stated that the exact location of an electron could not be determined. All measuring techniques would necessarily remove the electron from its normal environment. This uncertainty principle meant that only a population probability could be determined. Otherwise coincidence was the determining factor. Einstein did not want to accept this consequence ("God does not play dice"). Finally, Erwin Schrödinger formulated the electron wave function to describe this population space or probability density. This equation, particularly through the work of Max Born, led to the so-called "orbitals". These have a completely different appearance to the clear orbits of Bohr.

Chemists also adopted this concept and formulated such orbitals for chemical bonds and whole molecules. This led to a deeper knowledge and understanding and also provided a plausible explanation for many phenomena. But the price was a highly abstract formalism. It was not trivial, especially as theoreticians claimed that improved computer technology would allow chemistry to become a predictable subject matter. This "theorization" very soon reached its limits.

As is often the case, predictions were not always fulfilled. Theory can be helpful in the search for molecules that are useful (dyes), human-friendly (medications), or functional (materials), but cannot necessarily guarantee a successful outcome. The creativity and knowledge as well as the experience and intuition of the researcher are not replaceable.

New exciting paradigm shifts have changed the face of chemistry. For example, organic chemists in the past were more interested in how molecules **reacted** with each other; increasing investigations into complex life processes showed that it could be more interesting to discover how molecules **interact** with each other thereby forming functional systems with astonishing properties. This type of chemistry is still in its infancy. This is also true of the so-called "nanoworld". Particles of the order of a few nanometers cross the line from quantum to classical mechanics. Correspondingly surprising are the discontinuous changes in the properties. Inorganic chemistry is currently experiencing an equally great change. Who would have predicted that large production plants could provide cleaner and more sterile inorganic products than any pharmaceutical process? The latter requires a germ-free environment, whereas the production of semiconductors and microchips dictates that the work area must even be free of dust. High-purity elements, specific doping, and novel alloys exhibit exciting properties to allow innovative solutions to problems.

Even veteran elements can still present surprises. Many elements exist as different allotropes. This means that the atoms are arranged differently. In the case of carbon, the amorphous (soot), the dull gray graphite, and the brilliant diamond forms were known. It was therefore a great surprise when a new form was discovered in 1982; the fullerenes opened up a completely new area of chemistry. Hence it is not too far-fetching to deduce that further secrets lie buried in the elements, not to mention their compounds.

The periodic system of the elements is not a human invention. This ordering principle is rooted in the fundamental secrets of nature. Each element has its determined place and its specific identity.

The strengthening of this knowledge and understanding and the exploitation of these properties and combinations are the challenging tasks facing chemistry as an applied science to ensure a lasting improvement in the quality of life of the ever-increasing human population.

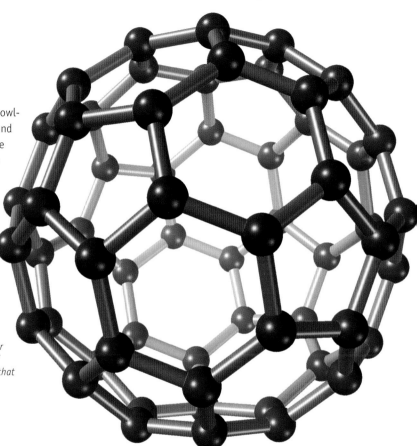

▶ The "Bucky ball" (fullerene or "football molecule") is a form of carbon first discovered in 1985 that enjoyed immediate popularity.

◀ *Whereas small and medium stars expand to red giants and at the end of their life emit a large portion of their mass as a gas cloud into space, larger stars explode in an unimaginable flash of light. All the energy that would normally be radiated over millions of years is — astronomically speaking — given off in a brief moment. The outburst is so immense that its light outshines entire galaxies. After the fire is quenched, a bizarre gas cloud remains that gradually expands in all directions. Its center contains an exotic entity that, depending on the original size of the star, could be a neutron star or even a black hole. Many processes occur under the extreme conditions of these events which lead to elements with higher atomic numbers. They form the "ashes", so to speak, that later shape celestial bodies such as planets.*
Picture: The Crab Nebula — remains of a supernova explosion observed by Chinese and Arabic astronomers in 1054 in the Taurus constellation.

The History of the Atomic Model

The question of the origin of the elements has been largely answered by astrophysics. After the Big Bang about 13 billion years ago, the universe expanded phenomenally, a process that continues today. The primordial (or original) mass consisted mainly of hydrogen as well as a certain amount of helium. Gravitation led to the formation of stars in which the matter was so highly compressed and heated that helium was formed from hydrogen by nuclear fusion. When the energy-releasing nuclear fusion had proceeded for a sufficient amount of time that the amount of hydrogen was on the decline, the star exploded to form a so-called red giant. Gravitation again attracted the cooling matter, which now consisted mainly of helium, and temperatures were reached that once again allowed nuclear fusion. This time, however, elements with higher atomic numbers, mainly carbon, were formed. This phase also ended with an explosion. The process repeated itself twice more, known as oxygen and silicon fusion. Elements with increasingly higher atomic numbers were formed. (The atomic number tells us the number of protons in the nucleus and hence the number of electrons in the atomic shell.) Thus the heavier elements were formed, and the Periodic Table completed itself. The amounts of the individual elements were very different. Furthermore, elements with even atomic numbers are more abundant than their neighbors with odd atomic numbers (Harkins's law).

When the Earth still consisted of stardust 4.5 billion years ago, it had undergone four such cycles. Hence each atom that we consist of carries a long history behind it. But why did it end? An explanation is provided by atomic physics. When the main component of the matter is the element iron, no energy is given off by further nuclear fusion. In contrast, it requires energy. This was shown in large particle accelerators in which nuclear fusion was studied. Furthermore, atoms whose atomic numbers are greater than 82, that of lead, become increasingly unstable. Then their disintegration (nuclear fission), gives off energy, as is the case with uranium in nuclear power stations.

The core of the Earth consists mainly of iron. It can be comically considered as a particle of ash that came to rest. However, this will not last forever. The sun is a star in its first phase. In 4 billion years, it will develop into a red giant, which will also envelop the Earth. But there is still a long time before then.

As well as we believe we know the past, so uncertain is the long-term prognosis. Will the universe continue to expand or will it reach an equilibrium state? Will the process of origin be reversed and end in a "Big Crunch"? This would essentially entail all matter being sucked into an inconceivable Black Hole. A new Big Bang could then take place. Would the same laws of nature come into play that we believe to have originated at and remained unchanged since the beginning? What are the actual origins of the laws of nature? Since Albert Einstein we know that if matter does not exist, then neither does time or space. What about mathematics? Would that be maintained? Questions about questions! Let us stay with a process that we can understand well: where does our knowledge of the atom come from?

The history of our concept of the atom is as exciting as a crime thriller, and for two reasons. First, there were only indications for the existence of the different atoms. This is still true today, as the pictures of the atom that have by now been seen by everyone are only visualized measurement data. There are no eyewitnesses. Second, there are no clues regarding a perpetrator or motive. Who created the atom and its laws as they are? Were the laws of nature already in place before the atom came into being, did they develop themselves or even change? And once again we find ourselves in metaphysics, where we — at least for now — really don't want to go.

As would be expected from such a fundamental problem, the root of the quest to unravel the composition of matter goes back to the pre-Socrates Greek philosophers. Even then the great minds pondered what made up nature and how it all came about.

▼ *In about 4 billion years our sun will also develop into a red giant. The diameter will then reach the orbit of Mars, and the inner planets will cease to exist.*

The Five Elements of Aristotle

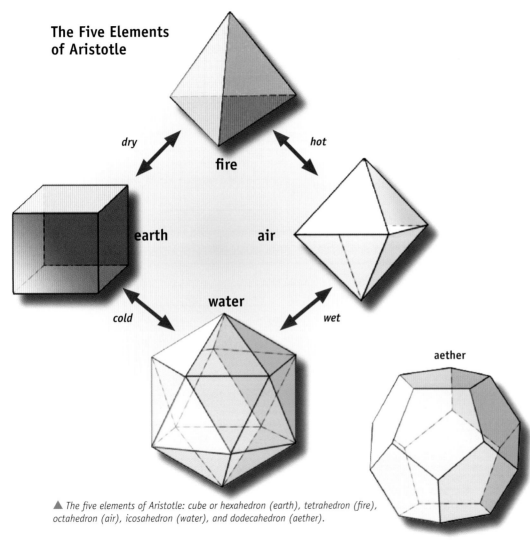

▲ The five elements of Aristotle: cube or hexahedron (earth), tetrahedron (fire), octahedron (air), icosahedron (water), and dodecahedron (aether).

The first food for thought was provided by Thales of Milet (ca. 625–547 BC). All life forms contain water. He quite plausibly concluded that water must have been the initial material. However, his hypothesis got off to a bad start, as his student Anaximander (611–546 VC, inventor of the sundial) did not believe it. He believed in an abstract "first principle" (the apeiron) that formed a cycle from which things arise and to which things fall. Anaximenes (ca. 585–525 BC) saw air as the initial material. For Pythagoras (570–500 BC), everything began with numbers. Quite an abstract concept; however, we should bear in mind that the Bible is also not much more concrete with its statement "In the beginning was the Word". The teachings of Empedokles (492–432 BC) were more tangible. He believed that the world consisted of four elements: water, earth, fire, and air. This concept provided many answers, and it is no surprise that it soon found many supporters. The most prominent were Plato (427–347 BC) and Aristotle (384–322 BC). They assigned one of the five platonic bodies to each element (cube for earth, octahedron for air, tetrahedron for fire, and icosahedron for water). The 12-faced dodecahedron represented the cosmos. The latter, according to Aristotle, was filled with aether. Alchemists in the Middle Ages occupied themselves with the search for this quintessential substance. Proof that this aether does not even exist could only be provided by Albert Einstein with this theory of relativity.

For now, let us remain in antiquity. A quiet, withdrawn, lateral thinker, Leukipp, pursued an imaginary experiment in 5 BC. It is known that matter can be divided and crushed. But how often can these processes be repeated? Certainly not infinitely. At some point, there must be an end, and one will encounter the indivisible, in Greek, à-tomos. Fortunately, Leukipp had a student, Democritus of Abdera (ca. 460–370 BC), an intelligent and avid communicator who combined a zest for life with a fondness of travel. From the hypothesis of the atom of his teacher, he built an entire area of thought, which he successfully disseminated among the people. All matter consists of atoms, which in turn are made up of the same matter. (Almost as if Leukipp and Democritus anticipated the existence of electrons, protons, and neutrons.) The differences in all substances arose from the shape, position, and arrangement of the atoms. Between the atoms was empty space. (This too proved to be correct.) As Democritus believed that the soul consisted of small "fire" atoms, he could be considered as a materialist. But in his ethics, he placed high standards on the human spirit. Despite his teachings on atoms, he believed that there was something else that stood above matter.

For almost two and a half millenia this question was laid to rest. No new facts came to light, and the biblical version of the Creation explained everything anyway. To question the secrets of God was heretical, even dangerous in the Middle Ages. The Inquisition did not even deliberate long over such people. Only during the Enlightenment was this ancient question once again raised, and eventually experimentally and theoretically answered.

During the production of the "Historical Periodic Table", for which this book was written as a supplement, it was difficult to do justice to the historical sequence of events and recognitions in atom research. The reason is trivial. With regard to the history of the idea of the atom (bottom row of the poster), for graphical reasons there was only place for 14 people (14 lanthanoids and actinoids). The difficulty lay in deciding whom to represent and whom to omit without doing injustice to anyone. The experts who were consulted answered these questions very differently, and at some point the decision had to be based on the best possible knowledge and belief. The choice made was primarily to awaken interest in this unique historical development. Particular significance was placed on methodological advances and paradigm shifts. I hope for the understanding of those who would have made different choices here and there.

Regarding the "ancestral lineage" on the poster, Democritus, of course, takes first place. Then comes John Dalton (1766–1844), an unassuming religious person whose broad scientific knowledge was self-taught. With his law of multiple proportions of the elements in their different chemical compounds, he rightfully saw indirect proof for the existence of atoms. He represented them as solid balls, to which he gave specific symbols. His precise concept of the element took this school teacher from a humble background to immortality.

An often undervalued contribution to the concept of the atom was that of Lorenzo Avogadro (1776–1856; 2006 was the 150th anniversary of his death). He came from a wealthy family and was a lawyer and philanthropist, but his real interest lay in science. Data, experiments, and astute consideration led him to the realization that equal volumes of gases must also contain the same number of atoms. As the atomic weights of some gases were known, he was able to determine the "atomic volume" to be 22.4 L at 20 °C. Furthermore, he recognized the fact that the elements oxygen and hydrogen existed as diatomic compounds. He therefore defined the term "molecule", which became a keyword in chemistry.

The picture gallery is conspicuous for the large gap between 1811 (Lorenzo Avogadro) and 1895

X-rays 1895	Radioactivity 1896	Electrons 1897	Quantum physics 1900	Isotopes 1900	Plum–pudding 1911	Planetary model 1913	Pauli principle 1924	Quantum mechanics 1926	Wave mechanics 1926	Statistical qua mechanics
Wilhelm C. Röntgen (1845–1923)	Antoine H. Becquerel (1852–1908)	Sir Joseph J. Thomson (1856–1940)	Max K. E. L. Planck (1858–1947)	Frederick Soddy (1877–1956)	Sir Ernest Rutherford (1871–1937)	Niels H. D. Bohr (1885–1962)	Wolfgang Pauli (1900–1958)	Werner K. Heisenberg (1901–1976)	Erwin Schrödinger (1887–1961)	Max Born (1882–1970)

(Wilhelm Röntgen). During this period, there were spectacular advances in all areas of science. In this context, the development of the periodic system by Julius Lothar Meyer and Dimitri Mendeleev was one of the most important steps. Nevertheless, the concept of the atom continued to be debated. Indeed, the debate began to intensify towards the end of the 19th century. Ardent supporters of atomism such as Ludwig Boltzmann (1844–1906) met with fierce resistance from, among others, Ernst Mach (1838–1916). For epistemological reasons he insisted on hypothesis-free science. As long as atoms could not be directly proved, their existence was to be doubted.

It was exactly this purism that was the guiding principle for the choice of researchers presented. This is already clear with the example of Wilhelm Röntgen (1845–1923; first Nobel Prize for physics 1901). Röntgen made no direct contribution to the atomic model. However, his discovery of the cathode ray (X-rays) provided a tool that allowed a glimpse into the atomic shell. That allowed the young Henry Moseley (1887–1915) to determine unambiguously the sequence of the elements according to atomic number.

The guiding principle that was decisive in the choice of Röntgen is also true for Antoine H. Becquerel (1852–1908; Nobel Prize for physics 1903 together with Pierre and Marie Curie). His discovery of radioactivity was not only the basis for the unraveling of new elements (radium and polonium by the Curies). Radioactivity and its phenomena became a universal tool that provided succeeding chemists and physicists with insight into the world of atoms.

Sir Joseph Thomson (1856–1940; Nobel Prize for physics 1906) discovered the free electron in 1897 and researched its properties. On the basis of the available knowledge, he developed a first concrete atomic model in which the electrons were embedded in a positively charged mass. Although his "plum–pudding model" was soon found to be incorrect, this in no way diminished the significance of this versatile and successful researcher.

Max Planck (1858–1947; Nobel Prize for physics 1918) at first did not have the atom in his sights. He was more interested in thermodynamics, and especially in the laws of radiation. In 1900 he surprised the Physical Society of Berlin — and later the whole world — with an experimentally based realization that changed the world view. In contrast to time and space, energy is quantized. Thus it does not form a continuum, but is essentially "grainy". The smallest unit is the Planck constant, a fundamental natural constant.

Frederic Soddy (1877–1956; Nobel Prize for chemistry 1921) together with Sir William Ramsay proved in 1903 that helium is formed in the disintegration of radium. In this he recognized a phenomenon of paramount importance and in 1913 he coined the term "isotope" for atoms with the same atomic number but different atomic weights as a result of different numbers of neutrons. The isotopes of an element exhibit identical chemical behavior, but under certain conditions have different physical properties (isotopic separation). As it later turned out, an increase in atomic number is accompanied by an increase in the ratio of neutrons to protons.

Sir Ernest Rutherford (1871–1937; Nobel Prize for chemistry 1908, which as a physicist he puzzled over) was a brilliant experimentalist endowed with an equal genius of being able to interpret the results. He recognized three types of radiation (alpha, beta, and gamma). He used scattering experiments with alpha radiation, which consists of helium nuclei, to prove that the atom is almost empty. The diameter of the atomic nucleus is about 10 000 times smaller than the atom itself. Furthermore, he proved that atoms are not indivisible and that in addition to protons, there must also be neutrons present in their nucleus. With Niels Bohr he developed the core–shell model of the atom.

Niels Bohr (1885–1962; Nobel Prize for physics 1922) became famous for his equally elegant, but nevertheless incorrect, atom model. In 1913 all facts seemed to indicate that the electrons did not buzz like bees around the nucleus.

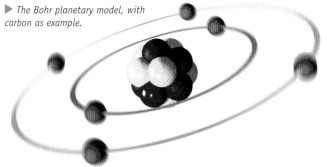

▶ *The Bohr planetary model, with carbon as example.*

Particularly spectra and quantum theory seemed to indicate an order. A planetary model almost suggested itself, but according to classical physics, the moving electrons should emit energy and consequently collapse into the nucleus. The 28-year-old Niels Bohr ignored this principle and postulated that the electrons in these orbits were "out of law". This clearly meant that classical physics could not describe or explain the properties of the atoms. The framework of physical theory came crashing down. Fundamentally new models had to be developed.[1]

[1] Didactically, the shell model has persisted till today owing to its clearness. Seven shells can be recognized (K, L, M, N, O, P, Q). Each of these has a certain number of orbitals (s, p, d, f). The s orbital can accommodate two electrons, the p orbital six, the d orbital 10, and the f orbital 14. The transition metals and the lanthanides and actinides have filled inner shells, which explains their chemical differences as well as their similarities. The subdivision into eight main groups and ten secondary groups was abandoned. According to international consensus, the Periodic Table today consists of 18 groups.

Graphical representation of the orbitals

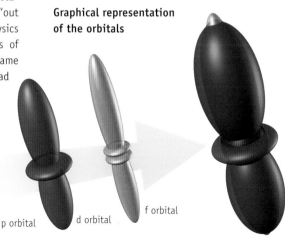

s orbital   p orbital   d orbital   f orbital

Werner Heisenberg (1901–1976; Nobel Prize for physics 1932) developed quantum mechanics, which allowed an accurate description of the atom. Together with his teacher and friend Niels Bohr, he elaborated the consequences in the "Copenhagen Interpretation" — a new world view. He found that the classical laws of physics are not valid at the atomic level. Coincidence and probability replaced cause and effect. According to the Heisenberg Uncertainty Principle, the location and momentum of atomic particles cannot be determined simultaneously. If the value of one is measured, the other is necessarily changed.

Erwin Schrödinger (1887–1961; Nobel Prize for physics 1932) transferred the concept of wave–particle duality of matter developed by L. V. de Broglie for electrons to the whole atom and thus developed wave mechanics. The Schrödinger equation allows a description of orbitals as the probability of the location of the electrons. Wave mechanics represented a significant development, but were subsequently shown to be insufficient.

Max Born (1882–1970; Nobel Prize for physics 1954) laid down the foundation for the further development of quantum theory in 1926 with his statistical interpretation of quantum mechanics. Above all, a theoretical interpretation of the chemical bond was possible. Max Born was one of the most significant theoretical physicists of the 20th century, a successful and versatile university lecturer (W. Heisenberg, P. Jordan, etc.), as well as a fierce advocate for the perception of scientific responsibility.

Wolfgang Pauli (1900–1958; Nobel Prize 1945), at the age of 24, formulated the exclusion principle, which became famous as the Pauli principle. Accordingly, all electrons in an atom differ from each other, none are the same. His theoretical considerations led him to the existence of so-called nuclear spins, which also explained the hyperfine structures of spectral lines. His hypothesis was later unambiguously confirmed. As each element has its own electron configuration, the different chemical properties could also be explained.

All the above-mentioned scientific contributions regarding the properties of the atom were pioneering efforts and made a lasting impression on chemistry. Furthermore, theoretical physicists have actively pursued the answer to the question as to what matter actually is. After the theory of relativity, matter could be related to energy by the famous equation $E = m \cdot c^2$. Other important question followed, for example, what forces keep the compact positive charges in the nucleus together? It came to light that, paradoxically, the strongest

▶ The search for a universal formula is not yet over.

forces are at play between the smallest particles. In contrast, the weakest force, gravitation, is involved between the largest bodies in the universe. The search for a universal formula that encompasses all four forces (gravitational forces, electromagnetic forces, weak nuclear forces between the constituents of the atomic nucleus that are responsible for beta decay, as well as the strong nuclear force in the atomic nucleus) in a single theory has not yet ended successfully. A new direction in the description of the relationship between matter and energy is the "String theory", according to which matter consists of compact, oscillating energy bundles (strings). At this subatomic level, the universe would be uniformly structured. These are the leading questions in natural philosophy.

Chemistry still has many concrete problems to solve. From the ocean of possible compounds — about 30 million are known and an end is not in sight — it is important to find those that are useful and valuable. With respect to the self-assembly of molecules, we are still at the early stages of recognition and application; this is especially true of molecular systems with functional properties. The production of materials with unusual or even still unknown properties is a promising field of research. And finally, chemistry and biology face the challenge to elucidate and understand complex life processes at the molecular level. We should not allow past achievements to deceive us into forgetting the long path and the many surprises that lie ahead.

▶ *Millions of substances are stored in the compound libraries of large firms.*

History of the Atomic Model

Democritus	(460–375 BC)	Idea of the atom	
John Dalton	(1766–1844)	First atomic model	1803
Lorenzo Avogadro	(1776–1856)	Atomic volume	1811
Wilhelm Röntgen	(1845–1923)	X-rays	1895
Antoine H. Becquerel	(1852–1908)	Radioactivity	1896
Sir Joseph Thomson	(1856–1940)	Electrons	1897
Max K.E.L. Planck	(1858–1947)	Quantum physics	1900
Frederick Soddy	(1877–1956)	Isotopes	1900
Sir Ernest Rutherford	(1871–1937)	Core-shell model	1911
Niels Bohr	(1885–1962)	Planetary model	1913
Wolfgang Pauli	(1900–1958)	Pauli exclusion principle	1924
Werner K. Heisenberg	(1901–1976)	Quantum mechanics	1926
Erwin Schrödinger	(1887–1961)	Wave mechanics	1926
Max Born	(1882–1970)	Statistical quantum mechanics	1926

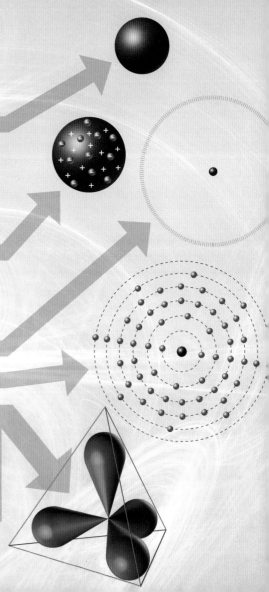

The Elements of the World
Properties, Discovery, Occurence, and Applications

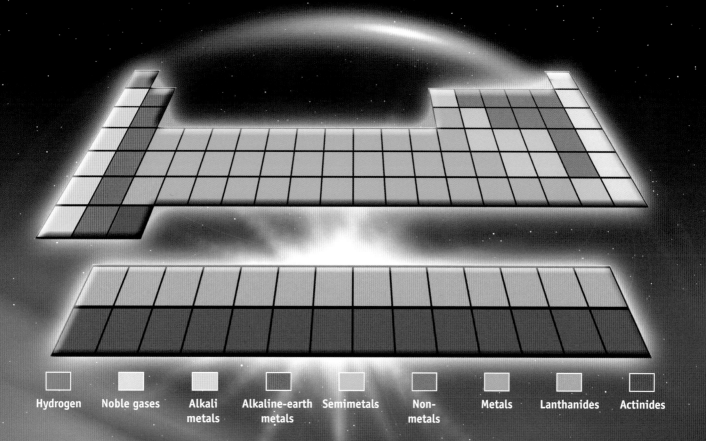

H 1
Hydrogen

Atomic weight
1.008

Isotopes
Deuterium
Tritium

1766 Henry Cavendish (1731–1810) recognized hydrogen as an element.

1661 Isolated by Robert Boyle (1627–1691) as "factitious air" upon treatment of metals with acids; previously observed by alchemists.

◀ In a water molecule, the hydrogen atoms are not located opposite each other, but form an angle of 105°. The dipole moment imparts special properties on the molecule, for example, the ability to form hydrogen bonds: O···H···O (20–40 kJ/mol⁻¹).

◀ Fuel cells show great promise as important future energy sources, especially in automobiles. Hydrogen reacts with oxygen from the air to form water, and electricity is generated in the process. The hydrogen is not necessarily obtained from crude oil (carbon, sun, etc.).

Name: from *hydro genes* (Greek = water-forming)

Properties

The most common element in the universe consists of one proton and one electron. In all stars, nuclear fusion generates helium. The small hydrogen atom (atomic radius: 0.078 nm) can be considered as the "Sunnyboy" of the elements. It is also the most common element in life forms; 60 % of the atoms in our bodies are hydrogen atoms. Hydrogen reacts violently with oxygen (detonating or oxyhydrogen gas) to form water. The unusual properties of water (high boiling point of 100 °C, ice being less dense than water, etc.) are due to the hydrogen bonds. These result from the electric attraction of the positively charged proton of one water molecule and the partially negatively charged oxygen atom of another water molecule (10–20 kJ mol⁻¹). These hydrogen-bonding interactions between oxygen and hydrogen atoms are essential to life (for example, they keep DNA and proteins together).

Only the isotopes of hydrogen have their own names: deuterium and tritium. "Naked" hydrogen, the proton, catalyzes many important reactions.

Hydrogen has widespread use: in ammonia synthesis and welding, as rocket fuel, reducing agent (e.g. for fats, desulfurization of oil products, etc.). H_2 is the promising fuel of the future: hydrogen engines, fuel cells, and maybe even nuclear fusion.

▶ Model of the DNA double helix. Hydrogen-bonding bridges keep the two DNA strands together. Without hydrogen bonds, life would not be possible.

He 2
Helium

Atomic weight
4.003

Noble gas

1895 Sir William Ramsay (1852–1916; Nobel Prize for chemistry 1904) isolated the rare gas.

1868 Observed in the spectrum of sunlight by Pierre Jules César Jansen (1824–1908) as well as by Norman Lockyer (1836–1920) and Edward Frankland (1825–1899). They believed the element to be a metal. Formally, they are the "discoverers".

Name: from *helios* (Greek = sun)

Properties
Discovered in the spectrum of the sun. Rare on Earth, small amounts are found in natural gas. The only element that has no solid phase under normal pressure; close to absolute zero it becomes a supercritical fluid. Less dense than air, non-explosive: ideal for balloons and airships. In the universe, helium is formed by nuclear fusion in the hottest zones; on earth it serves as a low-temperature cooling agent (boiling point: –268.9 °C = 4.2 K [record]). At these temperatures there is virtually chemical rigor mortis, but the conditions for superconductivity are fantastic. Hence liquid helium is used as a cooling agent when very strong magnetic fields are required (NMR, fusion experiments, etc.). Alpha radiation consists of "naked" helium cores (consisting of two protons and two neutrons).

Li 3
Lithium

Atomic weight
6.941

1817 Johan August Arfvedson (1792–1841) discovered this element during the accurate analysis of the mineral petalite.

1821 First isolated by William Thomas Brande (1788–1866).

▶ Lithium was a rarity for a long time. Microelectronics led to an increased demand for mini-batteries. Today, these little helpers are everywhere.

Name: from *lithos* (Greek = stone)

Properties
Can be found in small amounts almost everywhere. Soft element, the lightest solid element. Common in chemistry as a hydride. Organolithium compounds are important synthetic building blocks. Lithium became popular as an anode metal for powerful batteries as the lithium ion is small and mobile. These energy dispensers can be very small and provide power for pacemakers, hearing aids, etc. Lithium salts are employed in lubricants and in fireworks (red color). Lithium ions act against depression.

▶ *Early pacemaker.*

Be 4
Beryllium

Atomic weight
9.012

◀ Watch springs are made out of beryllium alloys.

1797 Louis-Nicolas Vauquelin (1763–1829) discovered the element in the semi-precious stone beryl.

1828 Described by Friedrich Wöhler as a metal.

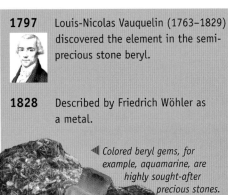

◀ Colored beryl gems, for example, aquamarine, are highly sought-after precious stones. They caught the eye of researchers and were thus discovered relatively early.

Name: derived from *beryl* (mineral)

Properties
A toxic element! Displaces magnesium in enzymes. Quite a soft metal, chemically relatively inert. In alloys with nickel and copper, even small amounts improve their electrical and thermal conductivity as well as their mechanical properties, for example, elasticity (application in watch springs). Beryllium is transparent to X-rays and can be used in "windows" in such applications. Is a component of the semiprecious stone beryl ($Be_3Al_2Si_6O_{18}$), whose colored variants, such as aquamarine, emerald, and red beryl are highly treasured and expensive. Colorless beryl (goshenite) was previously cut and polished to give lenses. Beryllium makes alloys hard, firm, corrosion-resistant, as well as nonsparking (tools, airplane brakes), and is used for precision instruments in space technology. It serves as a moderator in nuclear technology as it emits neutrons upon alpha radiation.

B 5
Boron

Atomic weight
10.81

1808 Joseph Louis Gay-Lussac (1778–1850) and Louis Jacques Thénard (1777–1857).

Obtained by Sir Humphry Davy (1778–1829) by reduction of boronic acid (H_3BO_3) with potassium.

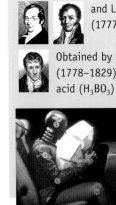

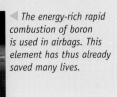

◀ The energy-rich rapid combustion of boron is used in airbags. This element has thus already saved many lives.

Name: fromn *burag* (Arabic = borax)

Properties
Dark brown metalloid that is extracted from the long-known borax (sodium borate).

Evolution did not use this element, only in certain plants is it important as a trace element. The element became well-known because of heat-resistant borosilicate glasses. Boranes are chemically interesting as B–H bonds react very specifically. Perborates are used in laundry detergents (Persil). The hardness of cubic boron nitride approaches that of diamond. Amorphous (brown) boron burns very quickly and gives off much heat and is therefore used in solid-propellant rockets and in igniters in airbags.

C 6
Carbon

Atomic weight
12.01

Carbon has been known as charcoal since early human history. It was identified as an element by Carl Wilhelm Scheele (1742–1786) and Antoine-Laurent Lavoisier (1743–1794).

◀ *Cavemen were already using charcoal 20 000 years ago in cave paintings. Carbon is still used for communication today in pencils and in printing ink.*

◀ *Bucky balls (football molecules) were only discovered in 1985 and named fullerene after the architect Buckminster-Fuller. The Nobel Prize for chemistry in 1996 was awarded for this new carbon chemistry. Molecular tubes with this structure have particularly interesting properties.*

▶ *Diamonds consist of pure, crystalline carbon (virtually single crystals). These artistically cut and brilliant stones pack a lot of value in the tiniest of volumes.*

Name: from *carbo* (Latin = coal); in some languages the word is also derived from *kol* (Indo-Germanic = to glow)

Properties

The element of life! Well-tolerated element, forms single and/or multiple bonds to itself and to its neighbors oxygen and nitrogen with infinite versatility. Hence its own branch of chemistry: organic chemistry. If there is life anywhere else in the universe, then it will with the greatest probability exist on the basis of carbon. The element is found in three contrasting allotropic modifications: graphite (soft), diamond (the hardest material), and fullerene (football molecules). The carbon sources coal, natural oil, and natural gas laid the foundations for modern civilization. However, increasing concentrations of CO_2 could lead to climate changes. Nature carefully recirculates the small amounts (ca. 380 ppm in air) in natural cycles. Small quantities (less than 1.4 %) of carbon convert iron into steel. Coke is used in blast furnaces to extract iron, soot is used as a pigment in printing ink, tires, and lacquers.

▶ *Petrochemical plant. Over 90 % of chemical products are based on natural oil that is mostly "cracked" to give simple products, which are laboriously separated by distillation. Of the 4 billion tons of natural oil consumed worldwide every year, only 7 % is used by the chemical industry.*

N 7
Nitrogen

Atomic weight
14.01

1772 Daniel Rutherford (1749–1819). The young doctor noticed that life forms and burning phosphorus only used part of air. The remainder he called "phlogisticated air", which we now know as nitrogen.

Carl Wilhelm Scheele (1742–1786) and Henry Cavendish (1731–1810) made similar observations at about the same time.

◀ Nodule-forming bacteria (legume bacteria) live in symbiosis with the root system of legumes (e.g. beans). They can reduce nitrogen to ammonia with the aid of a molybdenum–sulfur complex.

▼ Nitrogen fertilizers ensure plant growth. Retarding formulations prevent excessive washing away of the fertilizer from the soil.

Name: from *nitron genes* (Greek = saltpeter-forming). Antoine L. de Lavoisier (1743–1794) named the gas *azote* from *azoos* (Greek = lifeless)

Properties

Ubiquitous element, 78 % (volume) of the air is N_2. The strongest homoatomic bond (945 kJ mol^{-1}) is formed between the nitrogen atoms of N_2. Nitrogen is an inert gas, but nevertheless plays a key role in life processes. Some bacteria reduce N_2 to ammonia. This nitrogen fixation is not sufficient for intensive agriculture, and therefore ammonium nitrate is the main fertilizer. The catalytic reaction between nitrogen and hydrogen in the Haber–Bosch reaction went into large-scale production in Ludwigshafen in 1913. An example of a double-edged sword, ammonia, the "fertilizer from the air" ensures nutrition, but is also a precursor for explosives. The production of NH_3 is the largest chemical synthesis process worldwide. About 120 million tons per year ensure the sustenance of about 3 billion people. The product spectrum of ammonia is very broad; especially well-known is urea. Other prominent nitrogen compounds are amino acids, nitrogen oxides, laughing gas, and dynamite. Liquid nitrogen is a commonly used cooling agent (–194 °C).

▲ Modern ammonia plants produce about 1 million tons annually. The raw materials today are natural gas or crude oil fractions (coal was used earlier).

O 8
Oxygen

Atomic weight
16.00

1774 Carl Wilhelm Scheele (1742–1786) and Joseph Priestley (1733–1804).

Obtained upon heating HgO and was termed "fire air" or "dephlogisticated air".

▲ Combustion is the oldest application of a chemical reaction by humans (starting about 600 000 years ago). The burning process is the "motor" of civilization (ovens, pistons, turbines).

Name: from *oxy genes* (Greek = acid-forming), according to a definition of A. Lavoisier.

Properties

Colorless, reactive gas. Oxygen was not present in the initial atmosphere of the Earth, although at 50 % it is the most common element in the crust of the Earth (oxides, silicates, carbonates, etc.). The compound with hydrogen is remarkable. The hydrides of all other elements are unpleasant compounds, but H_2O is the molecule of life. The O_2 found in the air today, of which it makes up 20 %, was formed in the process of evolution by photosynthesis of algae, which then also allowed life on solid land. Oxidation with oxygen became and is still the dominant pathway of life forms for obtaining energy (respiration). Used in medicine in critical situations. Oxidations play a key role in chemistry (sulfuric acid, nitric acid, acetic acid, ethylene oxide, etc.). The ozone layer in space protects the Earth from cosmic UV radiation. Ozone (O_3) is used in the disinfection of water.

◀ As soon as humans leave their biosphere, they must take their oxygen supply with them. This is true underwater as well as in outer space.

▼ The atmosphere of the Earth is extremely thin. It evidently reacts much more strongly to change than initially thought.

F 9
Fluorine

Atomic weight
19.00

1886 Henri Moissan (1852–1907; Nobel Prize for chemistry 1906). Produced by electrolysis of potassium fluoride in aqueous hydrogen fluoride (HF).

1771 Carl Wilhelm Scheele (1742–1786) identified a unique acid (HF) in fluorspar.

◀ *Fluorine-containing polymers such as teflon are chemically and thermally particularly resistant. These surfaces also exhibit good nonstick properties, making them ideal materials in many areas in home and industry.*

Name: from *fluere* (Latin = flowing)

Properties
Georg Agricola (1495–1555) named fluorspar "fluor mineralis" (1529). the most aggressive and most reactive element with the highest electronegativity. Seizes an electron, and violently at that, from the outer shell of almost all other elements (even the noble gas xenon). Hydrofluoric acid (HF) even attacks glass (etching, matt light bulbs). Gaseous UF_6 is used in isotopic enrichment. Compounds with carbon have equally unusual properties (e.g. teflon). Chlorofluorocarbons (CFCs) found widespread use as a propellant and as a cooling agent. Owing to its stability, it transports chlorine into the stratosphere, where it damages the ozone layer. Hence its application was drastically limited. Life forms require fluorine as a trace element, for example, for hardening of teeth, hence it is a component of toothpaste.

◀ *Bones and teeth consist of hydroxylapatite (Ca phosphate). Tiny amounts of fluorine improves their resistance dramatically. Fluorine compounds in toothpaste prevent cavities.*

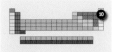

Ne 10
Neon

Atomic weight
20.18

1898 Sir William Ramsay (1852–1916) and Morris William Travers (1872–1961).

Predicted from the Periodic Table. Isolated by fractionation of argon and identified by spectral analysis.

Name: from *neos* (Greek = new); the noble gas group was unknown

Properties
Mendeleev could not predict the existence of the six noble gases. This group of monatomic gases is chemically inert — reserved and noble. No compounds of neon are known. Glow discharge of the element can bring it to luminesce (orange to scarlet red neon signs). Used in lasers. Present in the air: 18 ppm (volume). Separated by air liquefaction.

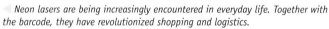

Neon lasers are being increasingly encountered in everyday life. Together with the barcode, they have revolutionized shopping and logistics.

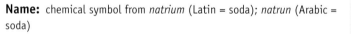

Na 11
Sodium

Atomic weight
22.99

1807 Sir Humphry Davy (1778–1829) isolated the element by electrolysis of molten caustic soda (NaOH).

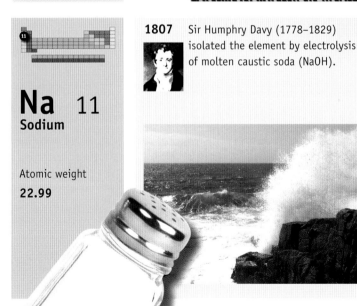

Name: chemical symbol from *natrium* (Latin = soda); *natrun* (Arabic = soda)

Properties
Soft, silvery metal, very reactive. Reacts vigorously with water and air, must be stored under paraffin oil. Used in industry as a strong reducing agent. Reacts with equally aggressive chlorine to form harmless salt known to be essential to life. As all life stemmed from the sea, all life forms require sodium ions, for example, for the conduction of the nerves and for humans to think. In humans (70 kg), 100 g of sodium can be found (as ions). Easy detection: makes flames yellow. Used in yellow lamps for street lighting. Sodium ions are widespread, for example, in glass, soap, mineral water, etc.

Sea water today contains about 35 g of NaCl per liter. In contrast, the physiological saline solution only has 9 g L^{-1}. When life was formed, the sea was not as salty.

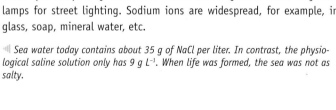

Mg 12
Magnesium

Atomic weight
24.30

1808 Sir Humphry Davy (1778–1829) obtained a small amount by electrolysis.

1831 Antoine-Alexandre-Brutus Bussy (1794–1882) obtained the metal by the reaction of magnesium chloride with potassium.

1755 Suspected to be an element by Joseph Black (1728–1799).

▶ Owing to its low specific weight, magnesium and its alloys are particularly suited for the construction of airplanes.

Name: from Magnesia, peninsula in Eastern Thessaly, where magnesite ($MgCO_3$) was found

Properties
Silver white, shiny, relatively soft metal. Used to make light alloys (airplanes). Pure magnesium burns vigorously (lycopodium), even underwater. The bright light is used in sparklers and pyrotechnics. Mg is found in abundance in dolomite ($MgCO_3 \bullet CaCO_3$) and sea water and is essential for all life forms. Interacts with Ca ions to control muscle contraction. Humans (70 kg) contain 19g and need 250–400 mg — a deficiency leads to cramps. Mg is the central atom in chlorophyll. In photosynthesis, this element controls the most important biochemical process on Earth: the production of about 200 billion tons of biomass per year. Magnesium is thus a key element for life.

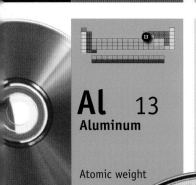

Al 13
Aluminum

Atomic weight
26.98

1827 Friedrich Wöhler (1800–1882) produced the element by reduction of aluminum chloride with potassium.

1825 Analytically recognized as an element by Hans Christian Oersted, but not isolated

◀ Mass requires more fuel. Hence, anything in cars that can be made from aluminum is made from this element.

Name: from *alumen* (Latin = alum)

Properties
Pure aluminum is soft, light, and malleable. It is the most common metal in the Earth's crust (ca. 8 %). Small amounts of Cu or Mg additives make it hard and firm. The surface is passivated with an oxide layer. Produced by fused-salt molten flux electrolysis. Cannot be welded, but is nevertheless optimal for airplanes (in which case it is riveted), construction units (windows, frames), and utensils such as cans, foil, and tubes. Increasingly found in cars in order to minimize weight. Tiny holes are burnt into extremely thin aluminum films in data-storage units. It has no function in physiology, but Al ions in the bloodstream can be detrimental.

Si 14
Silicon

Atomic weight
28.09

1824 Jöns Jakob Berzelius (1779–1848) treated potassium fluorosilicate with potassium to obtain amorphous silicon as a brown powder.

▶ *The production of memory chips requires highly pure silicon, which is grown as a single crystal (Czochralski process).*

▼ *Geothermal processes give rise to the most impressive form of silicon dioxide: rock (or mountain) crystal. This quartz form can also be worked into jewelry.*

Name: from *silex* (Latin = flint); flint commonly consists of quartz (SiO_2)

Properties

The highly pure crystalline form of this metalloid is the star attraction in the field of electronics. Without silicon, there would be no computers. It can be doped (addition of foreign metals) in a controlled manner and precisely etched with hydrofluoric acid (HF) for microchips. Silicon is the basic material in photovoltaics (electricity from sunlight). Forms very hard alloys (as SiC). The Si–C and Si–O bonds are stable: a good reason for the widespread use of silicones in medicine and technology. Silicates are the supporting substances in cement and glass. Some microorganisms in the sea form their skeletons from silicic acid; deposits provide kieselguhr with strong absorptive properties (for example, for nitroglycerine in dynamite). Silicon belongs to the same group as carbon, but, in contrast, cannot form stable double bonds. Life on the basis of silicon is hence very unlikely. The element is found in unlimited abundance, as sand on the beach (SiO_2). After oxygen, it is the second most common element in the Earth's crust.

◀ *The production of modern microprocessors poses the greatest technological and intellectual challenges. These "wafers" are the heart of all electronics and are still being developed.*

P 15
Phosphorus

Atomic weight
30.97

1669 Hennig Brand (ca. 1630–1692; left: idealized picture by J. Wright). Obtained from "sal microcosmicum" ($NaNH_4HPO_4$), which was isolated from urine.

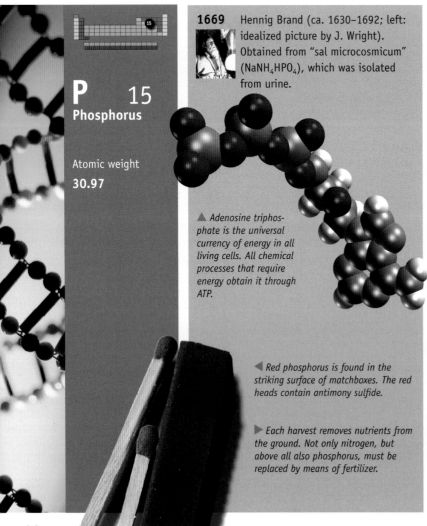

▲ Adenosine triphosphate is the universal currency of energy in all living cells. All chemical processes that require energy obtain it through ATP.

◀ Red phosphorus is found in the striking surface of matchboxes. The red heads contain antimony sulfide.

▶ Each harvest removes nutrients from the ground. Not only nitrogen, but above all also phosphorus, must be replaced by means of fertilizer.

Name: from *phosphoros* (Greek = light-bearer)

Properties

The only element that was discovered in body fluids (urine). This is plausible, as P plays a main role in all life processes. It is one of the five elements that make up DNA (besides C, H, N, and O; evolution did not require anything else to code all life). The P–O–P bond, phosphoric acid anhydride, is the universal energy currency in cells. The skeletons of mammals consists of Ca phosphate (hydroxylapatite). The element is encountered in several allotropic modifications: white phosphorus (soft, pyrophoric P_4, very toxic), red phosphorus (nontoxic, used to make the striking surface of matchboxes), black phosphorus (formed under high pressures). Phosphates are indispensable as fertilizer, but less desirable in washing agents as the waste water is too concentrated with this substance (eutrophication). It has a rich chemistry, is the basis for powerful insecticides, but also for warfare agents. A versatile element.

S 16
Sulfur

Atomic weight
32.07

Known from antiquity, assigned as an element in 1789 by Antoine-Laurent de Lavoisier (1743–1794).

Name: from *sulfurum* (Latin = sulfur) or *swep* (Sanskrit = sleep); "sulfur vapors" are poisonous

Properties

The "smelly shoe" of the elements. The oxidation product SO_2 has an acrid, burning smell, the reduction product H_2S stinks like rotten eggs and is very toxic. Sulfur is, nevertheless, a most useful element. It occurs in elemental form and has therefore been known for a long time: is mentioned in the Old Testament. Its main application today is in the production of fertilizers. Considerable amounts of sulfur are used in tire production for vulcanization. Sulfur is also a component of gunpowder. Physiologically indispensable as thioacetic acid and especially the S–S bridges that fix proteins in their shapes (e.g. insulin, but also in perms). A 70-kg human being contains 140 g of sulfur.

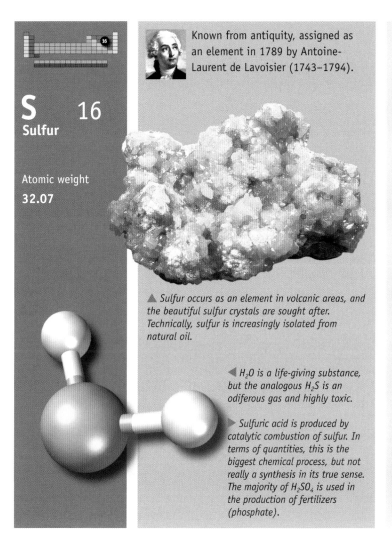

▲ Sulfur occurs as an element in volcanic areas, and the beautiful sulfur crystals are sought after. Technically, sulfur is increasingly isolated from natural oil.

◀ H_2O is a life-giving substance, but the analogous H_2S is an odiferous gas and highly toxic.

▶ Sulfuric acid is produced by catalytic combustion of sulfur. In terms of quantities, this is the biggest chemical process, but not really a synthesis in its true sense. The majority of H_2SO_4 is used in the production of fertilizers (phosphate).

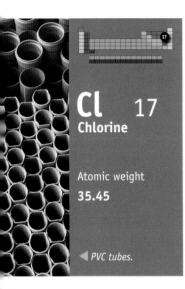

Cl 17
Chlorine

Atomic weight
35.45

◀ PVC tubes.

1774 Carl Wilhelm Scheele (1742–1786) discovered chlorine in the reaction of hydrochloric acid with manganese dioxide (pyrolusite or brownstone ore). Recognized as an element by Sir Humphry Davy (1778–1829).

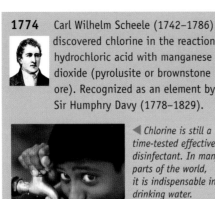

◀ Chlorine is still a time-tested effective disinfectant. In many parts of the world, it is indispensable in drinking water.

Name: from *chloros* (Greek = pale green)

Properties
The yellow green, pungent gas is toxic (used as a chemical warfare agent in World War I), but its bad reputation is undeserved. It is widely used as a disinfectant for drinking water and in swimming pools. It is one of the workhorses of chemistry. Many processes proceed via chlorine compounds as the element is very reactive. PVC is a particularly resistant plastic (important in the construction sector). Burning leads to hydrochloric acid, hence precautions must be taken. Many drugs contain chlorine, but there are also very toxic compounds, for example, dioxin. Can also be produced in nature as a result of fire, as plants also need chloride. There is about 35 g of salt (NaCl) dissolved in 1 L of sea water (physiological salt concentration: 9 g L^{-1}); A 70-kg human contains 95 g.

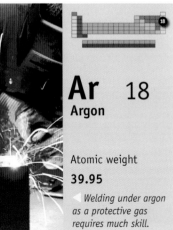

Ar 18
Argon

Atomic weight
39.95

◀ Welding under argon as a protective gas requires much skill.

1894 John William Lord Rayleigh (1842-1919; Nobel Prize for chemistry 1904) and Sir William Ramsay (1852-1916; Nobel Prize for physics 1904). Isolated by liquefaction of air and identified as a new element by spectral analysis.

Name: from *argos* (Greek = inert, inactive)

Properties
Nomen est omen. There are no known compounds of the colorless, odorless gas. At 1 %, there is almost 30 times more argon than CO_2 in the atmosphere. It accumulates during the liquefaction of air, and is a common and popular protective gas in chemistry and metallurgy, for example, in electrowelding. Ar is by far the most abundant noble gas. The green "argon line" is used in gas lasers. Light bulbs and discharge tubes also often contain argon.

▲ Argon lasers have many versatile applications in research and technology.

K 19
Potassium

Atomic weight
39.10

1807 Sir Humphry Davy (1778–1829) isolated the element through electrolysis of molten caustic potash (KOH).

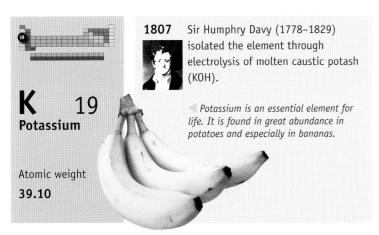

◀ *Potassium is an essential element for life. It is found in great abundance in potatoes and especially in bananas.*

Name: chemical symbol from *kalium* (Latin = potash) or *alkalja* (Arabic = ashes)

Properties

Soft, silvery metal, which is rapidly oxidized and reacts vigorously with water, with the formation of flames. Potassium ions are found in glass, in potash soap, and as nitrate in gunpowder and in matches. KCl is found in sea water and hence also occurs in salt domes. Most common application: as fertilizer. Potassium is essential to all life. It is removed from the ground during harvesting, and must be added again as it supports photosynthesis. Potassium is integrated in many bodily functions, for example, nerve conduction. Daily requirement: 2 to 3 g. Humans (70 kg) contain 140 g.

Ca 20
Calcium

Atomic weight
40.08

1808 Sir Humphry Davy (1778–1829) Obtained by electrolysis of a mixture of lime and silver oxide.

Name: from *calx* (Latin = limestone; calcium carbonate)

Properties

Silver white, relatively soft metal that is only applied in alloys. Oxygen and water attack pure Ca. The most prominent compound is the oxide (CaO) = burnt calcium, which hardens to calcium carbonate in mortar. Annual production of about 120 million tons. Burnt gypsum ($CaSO_4 \cdot 0.5\ H_2O$) hardens with water. A great step in evolution was the replacement of hard shells of brittle calcium carbonate by an internal skeleton of tough calcium phosphate (hydroxylapatite)–protein composite. Calcium is essential for all life forms. The daily requirement is 0.7–1.0 g. Humans (70 kg) contain 1 kg of calcium. Calcium silicate is the main component of cement. Marble is calcium carbonate in polycrystalline form and the favorite material of sculptors.

◀ *Shelled creatures such as mussels and snails are remarkable constructors of casings of Ca carbonate with protein layers.*

▶ *Marble is readily workable and has challenged artists throughout the ages.*

Sc 21
Scandium

Atomic weight
44.96

1879 Lars Frederik Nilson (1840–1899) found the element predicted by Dmitry Ivanovich Mendeleev (1834–1907) as "eka-boron" in the mineral gadolinite.

◁ Soviet engineers first used an aluminum–scandium alloy in the MiG 29.

Name: from *scandia* (Latin = Scandinavia)

Properties
Rare, silver white, soft metal that is almost as light as aluminum. Burns readily in air and reacts with water to form hydrogen. The element is actually not so rare, but is always present in small amounts in admixtures. It has relatively few applications, occasionally used in large television tubes. The iodide (ScI_3) contributes to the extreme brightness of mercury lamps (floodlights). The addition of scandium (2 %) to aluminum increases its strength. At first used in the Russian MiG 29, later also in racing bicycles, etc.

Ti 22
Titanium

Atomic weight
47.87

◁▷ *Titanium dioxide as a white pigment.*

1795 Martin Heinrich Klaproth (1743–1817). William Gregor (1761–1817), a minister who also dabbled in experiments, discovered a type of sand in 1791 from which a mysterious white powder could be produced. Klaproth recognized the yet-unknown element, and Henri Moissan (1852–1907) prepared pure titanium by electrolysis.

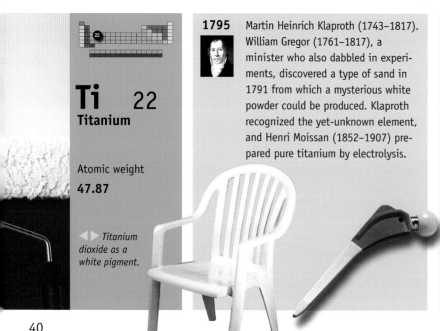

Name: from the titans (giants in Greek mythology)

Properties
Tough, shiny, silvery metal, protected from corrosion by a passivating oxide layer, but still maintains its matt shine. Important in the jewelry industry. Highly valued in the construction of chemical apparatus. Also highly suited for the construction of high-performance airplanes, as it is only 50 % heavier than aluminum (4.5 g cm^{-3}), but exhibits superior mechanical properties. Used in steel refinement. It is highly tolerated in the human body and hence commonly used for all types of implants. Its most common application, however, is as the oxide (TiO_2). The rutile form refracts light strongly and is therefore used as an exceptional white pigment in many applications, such as paint, lacquer, cosmetics, and even in toothpaste as it is completely inert. This makes titanium the "white giant" of the elements.

◁ *Titanium implants are widely used owing to its good mechanical properties and excellent compatibility. A hip joint ball is shown.*

V 23
Vanadium

Atomic weight
50.94

1831 Nils Gabriel Sefström (1787–1845) unambiguously proved its existence in Swedish bar iron.

1869 Henry Enfield Roscoe (1833–1915) produced the element by reduction of the chloride with hydrogen.

1801 Andres Manuel del Rio (1833–1849) suspected its presence in a Mexican mineral ore in 1801; but this was later shown to be vanadium.

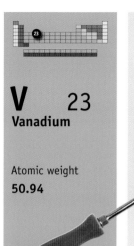

Name: derived from *Vanadis* (surname of the Nordic goddess of beauty and fertility Freyja), found in colored ores

Properties
Shiny silvery metal that is relatively soft in its pure form. Forms a highly resistant oxide coat. Used mainly in alloys, for example, in construction steel. Tiny amounts, in combination with other elements such as chromium, makes steel rustproof and improves its mechanical properties. Highly suited for tools and all types of machine parts. Also applied in airplane turbines. Chemically speaking, the element is of interest for catalysis (for example, removal of nitric oxides from waste gases). Vanadium forms countless beautiful, colored compounds (see **Name**). Essential for some organisms. Thus, natural oil, which was formed from marine life forms, contains substantial unwanted traces of vanadium that need to be removed.

Cr 24
Chromium

Atomic weight
52.00

1797 Nicolas Louis Vauquelin (1763–1829) discovers the element during the analysis of Siberian red lead ore ($PbCrO_4$).

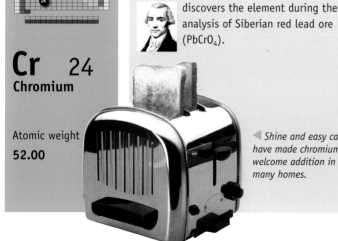

◀ *Shine and easy care have made chromium a welcome addition in many homes.*

Name: from *chroma* (Greek = color); forms many colored salts

Properties
The shining star of the elements. Hard metal, but ductile in its pure form, whose surface can be polished to a shine. Its main application is the alloying of heavy-duty steel (tools, drill bits). Chromium plating in all forms is highly popular. Even plastics can be chromium plated. Chromium dioxide is magnetic and is used in video tapes. Some chromates serve as pigments. Ruby owes its beautiful color to 0.25 % chromium. Chromium salts are used for tanning leather. Essential as a trace element for many organisms, including humans. Even available as a trace element in multivitamin preparatives (ca. 25 µg). However, hexavalent chromium salts are toxic (carcinogenic).

Mn 25
Manganese

Atomic weight
54.94

◀ After iron, manganese is the second-most-common heavy metal. In the manufacture of steel, it extracts sulfur and other impurities from iron. As an additive in alloys (2–25 %), it imparts a unique ductility and hardness to steel. Railway junction plates are prepared from these alloys.

1774 Johan Gottlieb Gahn (1745–1818) obtained metallic manganese from manganese dioxide ore (MnO_2), which had been known for a long time, by reduction with charcoal powder.

◀ Manganese ores are mined. On the ocean floor in the proximity of volcanoes there are so-called manganese nodules, which consist of oxides of manganese, iron and other heavy metals. They arise from deposition by microorganisms around a solid core (piece of mussel, shark tooth, etc.).

◀ Black dendrites in rocks are formed from manganese oxide depositions.

Name: from *magnes* (Latin = magnet); manganese dioxide (MnO_2) was for a long time thought to be some type of iron ore

Properties

Hard, brittle, silvery metal. Chemically not very resistant, but it imparts a unique ductility and hardness to steel, which makes it suitable for railway junction plates, tools, ploughshares, and knives. Manganese dioxide is suitable for batteries. Has an interesting chemistry as it can adopt oxidation states from −1 to +7. Potassium permanganate (deep violet) served for a long time as a disinfectant in medical applications. Essential as a trace element for all life, including plants. Humans (70 kg) contain about 12 mg, with a daily requirement of 0.5 to 1 µg (sources include wholewheat bread, wheatgerm, and nuts.

▶ Manganese–copper (–nickel) alloys are used as temperature-independent resistors. They are ubiquitous in electronics.

▼ Manganese dioxide (brownstone or pyrolusite) is used in dry batteries as an oxidizing agent for zinc (as counterelectrode).

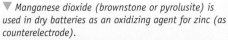

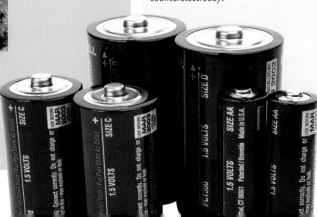

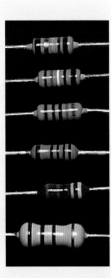

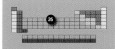

Fe 26
Iron

Atomic weight
55.85

1300 BC "Discovered" in the Middle East. The European "Iron Age" began in 750 BC. The alchemists assigned iron to Mars. Iron stands out above all other elements as determining the history of mankind owing to its use in tools and weapons.

Name: from *ferrum* (Latin); *isara* (Celtic); *iron* (Germanic)

Properties

Jack-of-(almost)-all-trades. Shiny, silvery, and soft in pure form. Tiny amounts of additives, such as carbon or other metals, change the properties dramatically. The element is magnetic (like cobalt and nickel) and makes up the core of the Earth. Iron is the most abundant of the heavy metals in the universe, as ^{56}Fe is the most stable atomic nucleus. This explains why exothermic nucleosynthesis ends at iron. Falls sporadically from the sky in meteors. Its relatively simple production (reduction of the ore with charcoal, later coke) makes it the most important and versatile material in civilization. A carbon content under 1.4 % makes iron a forgeable steel. World production of about 900 million tons per year. It is also essential in all forms of life. Warm-blooded animals, who have a high oxygen consumption, use iron as its transfer agent in hemoglobin. A 70-kg human carries 4.2 g of iron (just enough for a nail from which life hangs).

◀ *Modern casting techniques allow the production of massive steel rolls, which can subsequently be milled directly into plates for the car industry. Exact proportions of carbon and special additives make steel a "high-tech" material.*

▶ *Red blood cells owe their color to the iron complex heme, a component of hemoglobin. This provides the human body with oxygen, right up to the finest capillary. In the absence of oxygen, the blood goes dark red (venous blood).*

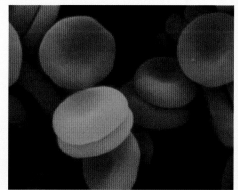

Co 27
Cobalt

Atomic weight **58.93**

1735 Georg Brandt (1694–1768) solved the puzzle of blue glass. In his dissertation he described the new element and finally shed light on the mysterious blue. Brandt was considered a great chemist, but nevertheless there is no picture of him.

◄ The beautiful "cobalt blue" in glazing was already known in ancient Egypt and is still popular today.

Name: derived from *Kobold* (German: malicious mountain troll, who hindered smelting, especially when the ore contained arsenic)

Properties

Silver blue, shiny, hard, particularly tough metal. Ferromagnetic, hence suitable for permanent magnets. Main application: particularly resilient alloys with iron, for example, razor blades, drill bits, tools, and jet engines. Famous/infamous is the isotope ^{60}Co, a strong gamma emitter. (Positive: tumor treatment; negative: long-term contamination from atomic weapons.) Surprisingly, cobalt is essential for most organisms. Humans require vitamin B12, whose central atom is cobalt; very little is required, but it is vital. A 70-kg human has only about 1.5 mg (a genuine trace element).

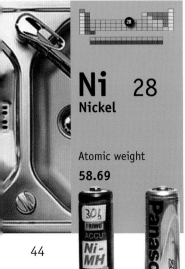

Ni 28
Nickel

Atomic weight **58.69**

1751 Axel Frederik Cronstedt (1722–1765) obtained this shiny element during his investigation of niccolite (NiAs).

◄ Not only the American nickel, but also many other coins contain this element.

Name: from *kupfernickel* (an older name for niccolite), named after "Old Nick" and his mischievous gnomes, because although it resembled copper ore, no copper could be extracted from it

Properties

Silver white, shiny, ferromagnetic, ductile metal. As it does not corrode, it is commonly used for nickel plating. Can, however, trigger allergies and therefore not a favorite for jewelry. Widely applied as an alloy in stainless steel, for example, for knives, utensils, airplane engines. Increasingly used in all sorts of accumulators and rechargeable batteries. Commonly used in coins: the American five-cent piece, the "nickel", contains up to 25 % of this metal. Important catalyst for hydrogenation (Raney-Ni). Essential trace element for many organisms, including humans.

Cu 29
Copper

Atomic weight
63.55

Copper was used in Europe from about 2000 till 1800 BC (Copper Age). Thereafter bronze was used; the alloy with about 20 % tin has superior properties.

◀ *The axe of "Ötzi the Iceman" (well-preserved mummy dated to 3300 BC) is made from copper.*

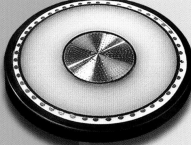

◀ *Cross section through a well insulated heavy-power cable (watertight).*

▶ *Breweries proudly display their old copper boilers, perhaps as a reminder of old traditions.*

◀ *Copper is the most commonly used coinage metal in the world. A further use: as a bactericide – a copper coin in a vase keeps flowers fresher for longer.*

Name: from *cuprum* (Latin = Cyprus); high-yielding deposits were found there

Properties

The reddish metal was already known in prehistoric times. It occasionally occurs as a native metal, but mostly in conspicuous green ores, from which it is extracted relatively easily. It is convenient to work, but not very hard. Not very optimal as a tool ("Ötzi the Iceman" had a copper axe with him). Only through the addition of tin is the more useful bronze obtained. Its zinc alloy is the versatile and widely used brass. Copper is one of the coinage metals. Water pipes are commonly made of copper. Its very good thermal and electrical conductivity is commonly exploited (cable!), as well as its durability (roofs, gutters), as the verdigris (basic copper carbonate) protects the metal. Cu phthalocyanines are the most beautiful blue pigments. Seems to be essential to all life as a trace element. In some molluscs, Cu replaces Fe in the heme complex. A 70-kg human contains 72 mg.

▶ *Brass (Cu–Zn alloy) is highly workable and has an impressive gold shine.*

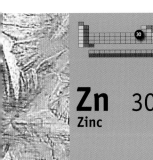

Zn 30
Zinc

Atomic weight
65.41

Metallic zinc was already known in ancient China. Was widely applied as brass, an alloy with copper.

1746 Andreas Sigismund Marggraf (1709–1782) first isolated the metal.

▶ *Zinc plating is still effective against rust. Forms impressive crystalline structures (like frost).*

Name: derived from *zinken* = jagged forms during the melting process; the term was introduced by Paracelsus (1494–1555)

Properties
Bluish, shimmering, brittle, relatively reactive metal. Is quite quickly covered with a protective oxide layer, which is why iron is treated with zinc With copper, forms the popular alloy brass, which was already known in antiquity. Used in batteries and as a stabilizer in plastics. Zinc oxide is used as a white pigment. Zinc ions are essential to all life forms, e.g., as a component of alcohol dehydrogenase and many other enzymes. Hence human beings (70 kg) carry about 2.3 g (half as much as iron).

▶ *The alloy with copper, brass, has widespread use.*

Ga 31
Gallium

Atomic weight
69.72

◀ UV LEDs

1875 Paul Emile Lecoq de Boisbaudran (1838–1912). Predicted by Dmitri Ivanovich Mendeleev (1834–1907) as "eka-aluminum"; was first identified spectroscopically in Spanish zincblende.

◀ *Gallium melts in the hand and then resembles mercury; however, it is not toxic. A gallium spoon melts in tea – a favorite trick.*

Name: from *gallia* (Latin = France); where it was discovered

Properties
Soft, silver white metal that melts in the hand (29.8 °C) and remains liquid up to 2204 °C (difference: 2174 °C, suitable for special thermometers). Gallium is quite widespread, but always in small amounts in admixtures. Its "career" took off with the advent of semiconductors. Ga arsenide and Ga phosphide, which are preferential to silicon in some applications, have extensive uses in microchips, diodes, lasers, and microwaves. The element is found in every mobile phone and computer. Ga nitride (GaN) is used in UV LEDs (ultraviolet light-emitting diodes). In this manner, a curiosity was transformed into a high-tech speciality.

Ge 32
Germanium

Atomic weight
72.64

1886 Clemens Winkler (1838–1904). Predicted by Dmitri Ivanovich Mendeleev (1834–1907) as "eka-silicon". Discovered during the analysis of argyrodite after a systematic search.

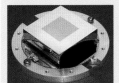

◀ *Germanium has maintained its importance in electronic through its use in rectifiers.*

Name: from *germania* (Latin = Germany); discovered in Freiburg

Properties
Germanium is a typical semimetal (metalloid). It is mostly found in zinc ores, is silver grey, very hard, and brittle. Its chemistry does not offer much of interest, but its physical properties are more fascinating. It occupies a prominent place as a semiconductor in transistors and rectifiers. Even though it is increasingly being replaced by other materials (silicon), the element can still be found in practically every electronic gadget. As it is transparent to infrared light, it is used as a special glass in the appropriate equipment (spectroscopy, IR visual display units). Its place of honor in electronics is secure as the first transistors were made of germanium crystals.

As 33
Arsenic

Atomic weight
74.92

As₂O₃

Already known in antiquity, as it occurs naturally in its elemental form.

1250 Isolated in its pure form by Albertus Magnus (1193–1280).

Name: from *arsenikon* (Greek = belonging to the male gender; the Greeks believed that elements had different genders

Properties
Although the element is a metalloid, the long, brittle, crystals have a metallic shine. The white, tasteless oxide (arsenic trioxide As_2O_3) has been famous and notorious ("inheritance powder"): even after centuries traces can be found in bodies. The arsenic compound "Salvarsan" was first used by Paul Ehrlich for the treatment of syphilis — the start of chemotherapy. Popular today as a semiconducting material. Component of LEDs (light-emitting diodes) and lasers. Arsenic hardens lead, used earlier in letterpress printing, today only for lead shot.

◀ *The semiconductor silicon obtains its remarkable properties upon doping with arsenic. Hence, practically all electronic devices contain this element.*

Se 34
Selenium

Atomic weight
78.96

1817 Jöns Jakob Berzelius (1779–1848) discovered the element in the sludge of the lead chamber in the sulfuric acid factory in Gripsholm.

▶ *Light-exposure meter based on selenium; upon illumination, the conductivity can increase by a factor of 1000.*

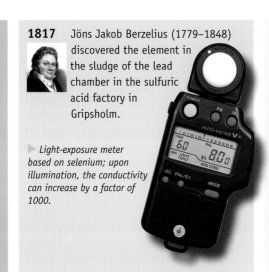

Name: derived from *Selene* (Greek goddess of the moon; was meant to show the close relationship to the previously discovered tellurium (from tellus (Latin = Earth))

Properties
As a true semimetal, there is a metallic as well as a nonmetallic form, which is made up of Se_8 rings (red, amorphous powder). Its chemistry is less spectacular than its physics (photosemiconductor, whose conductivity changes upon exposure to light). Light quanta result in the emission of electrons, hence ideal for photocells (light-exposure meters), electrophotography, copiers, and solar cells. Also useful for semiconductors and lasers. Good future prospects. Big surprise: selenium is toxic, but essential as a trace element for many organisms, including humans. A selenium deficiency can make one sick, but this is very rare. In humans (70 kg), there are 7 mg.

Br 35
Bromine

Atomic weight
79.90

1825 Antoine Jérôme Balard (1802–1876) treated a caustic solution obtained from the ashes of seaweed with chlorine water. Besides iodine, bromine was formed.
Discovered independently by Carl Löwig (1803–1890) in Heidelberg.

◀ *For almost 150 years, silver bromide layers on glass or plastic were the mainstay of photography. This era is now coming to an end.*

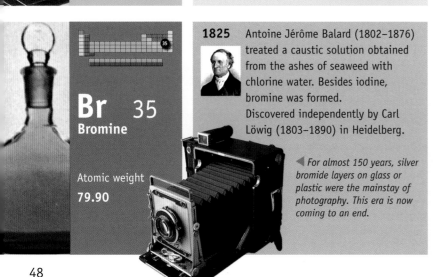

Name: from *bromos* (Greek = stench)

Properties
The only liquid nonmetal does full justice to its name and to its group (the halogens): it stinks and is aggressive. Bromine compounds capture radicals hence the application as flame retardant. Is a component of the previously widespread sleeping agent "Bromural"; as a consequence the element became popular (bromine calms). Its most important role is as a silver salt in photography. It is present in small amounts in our bodies (ca. 250 mg), but its function has still not been established.

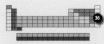

Kr 36
Krypton

Atomic weight
83.80

1898 Sir William Ramsay (1852–1916) and Morris William Travers (1872–1961) discovered it spectroscopically in the low-temperature fractionation of crude argon.

Besides its application in high-quality light bulbs, krypton is being increasingly used in the production of powerful lasers.

Name: from *kryptos* (Greek = hidden)

Properties
Rare noble gas. In 1 m³ of air there is only 1 cm³ of krypton (1:1 million). The freezing point (−157.4 °C) and boiling point (−153.2 °C) lie close to each other. Valuable by-product of air liquefaction, as it is used to fill high-quality light bulbs, especially energy-saving light bulbs, fluorescent lamps, and photographic flashes. The low thermal conductivity allows higher temperatures of the glow filaments. Some not very stable fluorine compounds have actually been prepared. UV lasers also use krypton.

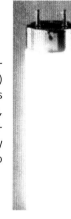

Rb 37
Rubidium

Atomic weight
85.47

1861 Robert Wilhelm Bunsen (1811–1899) and Gustav Robert Kirchhoff (1824–1887) discovered it in the mineral lepidolite through spectral analysis.

Transistors have replaced amplifier valves in niche applications. The function of this element is to bind traces of oxygen in the vacuum tube.

Name: from *rubidius* (Latin = dark red)

Properties
The element revealed itself through spectacular violet-colored flames and several red spectral lines. The metal melts at 38 °C, is very soft, and extremely reactive (burns in air and reacts violently with water). Rubidium is stored under mineral oil. It is suitable as a scavenger (oxygen capture) in vacuum tubes, where it is deposited on the glass as a mirror. It can also be found in photocells and phosphors for screens (for example, for air-traffic controllers). Not physiologically important. The radioactive rubidium-87 is useful for age determination in geochronology (half-life: ca. 50 billion years).

Sr 38
Strontium

Atomic weight
87.62

1808 Sir Humphry Davy (1778–1829) first succeeded in the production of the metal through electrolysis.

1790 Adair Crawford (1748-1795) first distinguished strontium ore (strontianite) from barite and other barium ores.

◂ Strontium-90 is a fission product of uranium, which is used in permanent nuclear batteries as an energy-rich beta emitter.

Name: from *Strontian* (the place in Scotland where the ore was discovered)

Properties
Silver white, soft metal that protects itself immediately with an oxide layer. Admired in all fireworks as red flares. Handy in emergencies (flare guns) to call for help. Strontium has acquired a bad reputation. The radioactive isotope ^{90}Sr is formed as "fall-out" in atomic tests and is incorporated instead of calcium in bones, especially in children. Justified protests led to the suspension of terrestrial atom tests. In this way an isotope changed world politics. The Nobel Laureate (chemistry) Linus Pauling was awarded the Nobel Peace Prize in 1962 for his involvement. ^{90}Sr is a strong beta-emitter that is used in nuclear medicine and as a nuclear battery, for example, in buoys. The element is found in many permanent magnets.

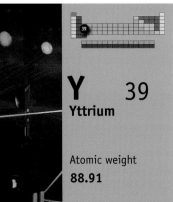

Y 39
Yttrium

Atomic weight
88.91

1794 Johan Gadolin (1760–1852) discovered the ore yttria in the Swedish village Ytterby.

1843 Carl Gustav Mosander isolated yttrium from the ore.

◂ Ferromagnetic yttrium–iron garnet, YIG, is used in radar devices. Its resonance frequency can be influenced by a magnetic field.

Name: from *Ytterby*, the place in Sweden where the corresponding ore was found

Properties
Soft, silvery metal that rapidly forms a protective oxide layer. Chemically speaking, the element is not very exciting. Electronics, lasers, and strong magnets, however, make it a rising star in solid-phase physics. The yttrium–aluminum garnet (YAG) laser is the workhorse of laser technology (cutting, welding, drilling). Furthermore, it is an important component of high-temperature superconductors (lose their electrical resistance at −180 °C). Yttrium oxide forms particularly fire-resistant ceramics. The oxide sulfite together with europium provides the red color on computer monitors. One should keep a close eye on this element, if nothing else because diamonds can be imitated by yttrium oxide.

Zr 40
Zirconium

Atomic weight
91.22

1789 Martin Heinrich Klaproth (1743–1817) isolated zirconium ore (ZrO_2) from the semi-precious stone hyacinth.

▶ *Doped zirconium dioxide is the solid electrolyte in lambda sensors (oxygen sensors used in the field of environmental protection).*

1824 Jöns Jakob Berzelius (1779–1848) produced the element by heating a mixture of potassium and potassium zirconium fluoride.

Name: from *zargun* (Arabic = gold color), a reference to the semiprecious stone zircon.

Properties
Hard, silvery, shiny metal that is extremely corrosion-resistant owing to an oxide layer. Used in alloys in nuclear technology (shells for fuel rods as it does not absorb neutrons) as well as in chemical apparatus (spinning jets, pump parts). The oxide ZrO_2 (zirconium dioxide) doped with Y_2O_3 is particularly important as a solid ion conductor in lambda sensors for the measurement of oxygen resides in exhaust gases. Zirconium oxide is highly fireproof (ceramics), extremely hard (abrasives), and has a high refractive index as a pure crystal. In the area of brilliant cut, zircon is almost as brilliant as diamond, but is not quite as hard.

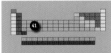

Nb 41
Niobium

Atomic weight
92.91

1801 Charles Hatchett (1765–1847). In 1846 Heinrich Rose (1795–1864) proved that Hatchett had obtained a mixture of two elements, which he separated into tantalum and niobium.

▲ *As little as 1 % of niobium makes steel a high-performance material for the car industry and railway vehicles.*

Name: derived from *Niobe*, daughter of Tanatalus (Greek mythology); niobium and tantalum always occur together

Properties
The bright, silvery, pure metal is relatively soft. Its main application is as an additive for rust-proof, heat-resistant steels that are used to manufacture cutting tools. HSLA (high-strength, low-alloy) steels contain less than 1 % niobium. They are used in the car industry and for pipelines. Higher doping levels increase the heat resistance of the alloy (turbine blades and rocket nozzles). Niobium alloys can be welded, hence it is one of the best enrichment agents for steel. Niobium–tungsten thermocouples can measure temperatures precisely up to 2000 °C. In contrast, the alloy Nb_3Sn is a prized low-temperature superconductor for extremely strong magnets (CERN).

▶ *Rocket engine*

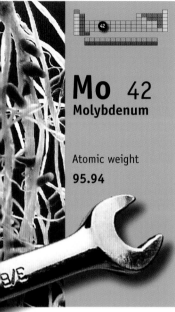

Mo 42
Molybdenum

Atomic weight
95.94

1778 Carl Wilhelm Scheele (1742–1786) isolated "molybdic acid".

1781 Peter Jacob Hjelm (1746–1813) succeeded in isolating the element by reduction of molybdenum oxide with carbon.

◀ Molybdenum alloys and coatings make steel particularly hardwearing for use in heavy-duty parts.

Name: from *molybdos* (Greek = lead)

Properties

The naming is suggestive of the fact that its specific weight resembles that of lead. That is where the similarity stops, as molybdenum is a high-strength, tough, and hard metal with a high boiling point (2623 °C). A small amount of this additive improves steel to a remarkable quality. The element is indispensable for high-performance steel, including steel for the automobile industry. The temperature resistance of the alloys allows their use in filaments and airplane and rocket parts. The sulfide is known for its excellent lubricant property, similar to that of graphite. Molybdenum is an essential trace element for all organisms. Molybdenum–sulfur complexes allow nitrogen fixation in certain bacteria.

▶ Molybdenum is not only excellent for the enrichment of steel, but the sulfide (MoS_2) also exhibits excellent lubricant properties.

Tc 43
Technetium

Atomic weight
98.91

1937 Emilio Gino Segrè (1905–1989) and Carlo Perrier (1886–1948). Ernest O. Lawrence (1901–1958) bombarded molybdenum with cyclotron-accelerated deuterons; element 43 could be proved in this experiment. Only radioactive isotopes exist.

Name: from *technikos* (Greek = artificial)

Properties

The metal is radioactive and does not occur in nature, as the half-life of all isotopes is shorter than 5 million years. It is found in readily isolable amounts in nuclear reactors. It is an effective "rust-preventer" for iron and steel in special applications. The metastable isotope ^{99}Tc has a half-life of only 6 hours and is therefore used as a gamma radiator in medicine (radiation therapy and diagnostics). Of very little commercial importance.

Ru 44
Ruthenium

Atomic weight
101.1

1844 Carl Ernst Claus (1796–1864) discovered the element.

1828 Gottfried Wilhelm Osann (1797–1866) suspected its existence in platinum residues.

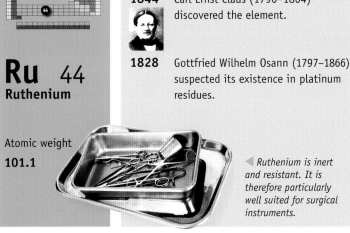

◀ Ruthenium is inert and resistant. It is therefore particularly well suited for surgical instruments.

Name: from *ruthenia* (Latin = Russia); discovered in Russian ore at the University of Kazan

▶ High-quality shaver foil for electric razors.

Properties
Rare, shiny, and lightest metal of the platinum group. Hardens platinum and palladium. The presence of 0.1 % of ruthenium in titanium improves its resistance to corrosion 100-fold. The spectacular catalytic properties of ruthenium are used on industrial scales (hydrogenations, sometimes enantioselective, and metathesis). Titanium electrodes coated with ruthenium oxide are applied in chlorine-alkaline electrolysis. Suitable for corrosion-resistant contacts and surgical instruments.

▶ Ruthenium is a versatile catalyst for industrial-scale chemical processes.

Rh 45
Rhodium

Atomic weight
102.9

1803 William Hyde Wollaston (1766–1828) found rhodium in crude platinum.

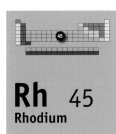

◀ Hardness and chemical resistance make rhodium a suitable metal for heavy-duty spinning nozzles for plastic fibers.

Name: from *rhodon* (Greek = rose); some rhodium salts are pink

Properties
The bluish white, hard, yet ductile, metal is inert to all acids and highly non-abrasive. Used for heavy-duty parts in electrical contacts and spinning jets. Reflectors are prepared from the mirror-smooth surfaces (e.g. head mirrors in medicine). Thin coatings provide a corrosion-resistant protective layer, for example, for jewelry, watches, and spectacle frames. The metal is a constituent of three-way catalysts. Rhodium complexes are used with great success in carbonylations (reactions with CO) and oxidations (nitric acid) in industry. Platinum–rhodium alloys are suitable thermocouples.

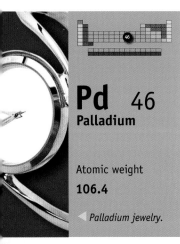

Pd 46
Palladium

Atomic weight
106.4

◀ *Palladium jewelry.*

1803 William Hyde Wollaston (1766–1828) also found palladium in crude platinum.

▲ *Palladium is an efficient and sought-after exhaust catalytic converter.*

Name: derived from *Pallas*, an asteroid discovered shortly before (1802).

Properties
Silver white, shiny metal that is ductile and readily worked into sheets. The metal is highly corrosion-resistant. Used for electrical contacts and surgical instruments. Attractive for jewelry, as gold–palladium alloys form the highly desirable white gold, not only for jewelry, but also for dental crowns. Palladium has the unique ability to absorb 900 times its weight or 3000 times its volume of hydrogen, thus making it the best hydrogenation catalyst. Palladium films are permeable to hydrogen and therefore suitable for fuel cells.

Ag 47
Silver

Atomic weight
107.9

Occurs naturally in its metallic form and was therefore already used in antiquity for jewelry and coins. Alchemists associated silver with the moon.

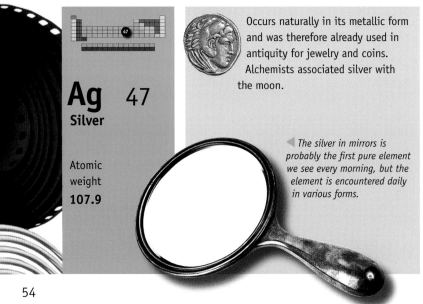

◀ *The silver in mirrors is probably the first pure element we see every morning, but the element is encountered daily in various forms.*

Name: from *siolfur* (Germanic); *silabar* (Old High German); *argentum* (Latin)

Properties
After gold and copper the third metal in the history of civilization. As it is easy to work, it was first highly prized for jewelry. Silver coins have been important from the invention of money until today. To date, approximately 1 million tons of silver have been extracted; another 15 000 tons are obtained annually. The element has the best electrical and thermal conductivity as well as the highest reflecting power (mirrors) in the Periodic Table. Light-sensitive silver chloride and bromide have been imprinted in the history of film and photography for 150 years. This era is now slowly coming to an end with the advent of optical data storage. It has a place of honor in good batteries, and still highly valued for jewelry and cutlery. Silver remains an excellent catalyst for oxidations (formaldehyde, ethylene oxide) in the chemical industry.

Cd 48
Cadmium

Atomic weight
112.4

1817 Friedrich Stromeyer (1776–1835). During the inspection of a pharmacy, Stromeyer confiscated zinc carbonate as it did not meet the purity standards; when calcined, it showed a yellow residue: cadmium oxide.

◀ *Cadmium-113 captures neutrons effectively and is suitable in the regulation of nuclear reactors. The isotope is a beta-emitter with a half-life of 10^{16} years (record).*

Name: from *cadmia* (Latin = calamine), a zinc ore in which cadmium occurs

Properties
A silver white, soft metal that shows a characteristic tarnishing upon exposure to air. The layer offers protection against corrosion. It was first popular as the striking cadmium yellow pigment (CdS), then used intensively in Ni–Cd rechargeable batteries. A pollution incident in Japan uncovered a biological hazard: cadmium displaces zinc in important proteins. Can lead to serious health problems and is no longer used to produce utility goods. Cadmium rods are used as regulators in nuclear plants owing to their efficient ability to capture thermal neutrons. Still used as an additive in alloys and in optoelectronics.

In 49
Indium

Atomic weight
114.8

1863 Ferdinand Reich (1799–1882) and Hieronymus Theodor Richter (1824–1898). Discovered in zincblende and spectroscopically identified as a new element.

◀ *ITO (indium tin oxide) is a transparent electrical conductor that is highly suited for touch screens.*

Name: from the indigo-colored bands in the line spectrum; makes flames blue violet

Properties
The low-melting-point (157 °C), silver metal is mainly used in alloys to decrease the melting point. Combined with tin, lead, and bismuth to produce soldering metal for wide temperature ranges. The element is highly valuable in the electronics age as its unique properties are ideal for solar cells, optoelectronics, and microwave equipment. The arsenide is used in lasers and is also suitable for transistors. ITO (indium tin oxide) is a transparent semiconductor with wide application in displays, touchscreens, etc. In the household, indium as an additive prevents the tarnishing of silverware. Some electronic wristwatches contain indium batteries.

Sn 50
Tin

Atomic weight
118.7

Its alloy with copper provided the material for the Bronze Age. Alchemists associated tin with the planet Jupiter.

▼ *Tin baths are becoming increasingly important for the soldering of electronic components with printed circuits.*

Name: from *zein* (Germanic = rod), the metal was cast in this shape; chemical symbol from *stannum* (Latin)

Properties
Soft, malleable metal whose characteristic gray color results from the oxide layer. Used as an alloy with copper since antiquity (Bronze Age): weapons, coins, jewelry, statues, bells, organ pipes. Tin was widely used for household utensils in the Middle Ages. Napoleon launched a competition to find a way to make food last longer. This led to the discovery of tinplate (tin-plated or galvanized iron). Ironically the buttons on Napoleonic uniforms were made of tin and fell victim to tin plague in the cold Russian winter of 1812 (conversion of tetragonal (beta) tin into cubic (alpha). Commonly used today to solder switches. Good stabilizer for plastics. Applied in ITO (see indium). Tin has 10 stable isotopes (record!).

Sb 51
Antimony

Atomic weight
121.8

◂ *Match heads contain antimony sulfide.*

Antimony was already used in antiquity and was recognized as an element at the beginning of the 17th century. It imparts a yellow orange color to glazing.

◂ *The success of print would not have been possible without type metal (Pb–Sn–Sb alloy) with 3–25 % antimony.*

Name: from *anti monos* (Greek = not alone); *stibium* (Latin = pen); black shiny antimonite/stibnite (Sb_2S_3) was used as an eyebrow pencil

Properties
Semimetal that occurs as a tin-type, brittle form and as a yellow, unstable, nonmetallic form. Its main use is in alloys to harden other metals. Without the addition of antimony, lead would have remained the "softy" of the Periodic Table. But with antimony, lead ruled the print world and later found use in the production of rechargeable batteries. It can be found in older ceramic glazing (yellow orange). Everyday encounters: antimony sulfide in match heads and red rubber, antimony oxide is used as a flame retardant. Pure antimony is starting to become of interest in the electronics sector.

Te 52
Tellurium

Atomic weight
127.6

1783 Franz Joseph Müller von Reichenstein (1740–1825).
The element was first thought to be antimony; identified as the element tellurium in 1798 by Heinrich Klaproth (1743–1817).

◀ Infrared cameras could not function without tellurium. They are used to detect heat loss; this is of importance in the renovation of old buildings.

Name: derived from *Tellus* (Roman goddess of the Earth)

Properties

The silver white, shiny, metal-like semiconductor is considered a semimetal. The atomic weight is greater than that of the following neighbor (iodine), because tellurium isotopes are neutron-rich (compare Ar/K). Its main use is in alloys, as the addition of small amounts considerably improves properties such as hardness and corrosion resistance. New applications of tellurium include optoelectronics (lasers), electrical resistors, thermoelectric elements (a current gives rise to a temperature gradient), photocopier drums, infrared cameras, and solar cells. Tellurium accelerates the vulcanization of rubber.

▶ Light-sensitive drums in laser printers are reliant on tellurium.

I 53
Iodine

Atomic weight
126.9

1811 Bernard Courtois (1777–1838).
When sulfuric acid was added to the ashes obtained from seaweed, a violet gas was given off that condensed as dark crystals with a metallic luster.

▲ To divert storms, clouds can be seeded with silver iodide crystals to promote "targeted" rain.

Name: from *iodos* (Greek = violet)

Properties

The black, shiny nonmetal sublimes easily to form a violet gas. The halogen is clearly less aggressive than its lighter relatives. In the early 19th century, its antiseptic properties were recognized and are still exploited today (iodine PVP). Then came the application of silver iodide as a sensitizer on photographic plates. Iodine is used as a promoter in many catalytic processes. It is used in halogen lamps. The connection between iodine deficiency and goiter was discovered in Switzerland. Like many other organisms, humans also require iodine (0.1 to 0.2 μg per day) for the thyroid hormone thyroxin. Iodine is the heaviest of the essential trace elements. A 70-kg human contains 12 to 20 mg.

▶ Disinfection

Xe 54
Xenon

Atomic weight
131.3

◀ *Sunbeds have come into fashion.*

1898 Sir William Ramsay (1852–1916) and Morris William Travers (1872–1961). Enriched by fractional distillation of krypton and identified spectroscopically as a new element.

▼ *Powerful xenon lights offer automobile drivers distinctly more vision and safety.*

Name: from *xenos* (Greek = stranger)

Properties
The rarest of the noble gases is also the one of which several compounds are known. The relatively stable fluorine compounds are good fluorinating agents. The chlorides, oxides, and oxifluorides decompose readily, sometimes even explosively. Its application is limited by its high cost. Intense, high-pressure lamps (floodlights) and tubes for sunbeds are filled with xenon. Long used in flashlights, more recently also in automobile headlights. Also applied in powerful lasers and in medicinal diagnostics (radioactive isotope ^{133}Xe). A mixture of 80 % xenon and 20 % oxygen is a well-tolerated anesthetic.

Cs 55
Cesium

Atomic weight
132.9

◀ *Atomic clock*

1860 Robert Wilhelm Bunsen (1811–1899) und Gustav Robert Kirchhoff (1824–1887) discovered this element during the spectral analysis of Dürkheim mineral water which showed two new blue lines.

◀ *The high susceptibility for photons makes cesium indispensable in sensitive photocells.*

Name: from *caesius* (Latin = sky blue)

Properties
Soft, golden, shiny metal that reacts rapidly with oxygen and explosively with water. A remarkable element: it has the largest atomic radius, is the most reactive metal, forms the strongest base (CsOH), melts at a low temperature (28.4 °C), and gives rise to salt solutions with a high specific gravity (centrifuge solutions). It has an unusual property in that it acts as a strong promoter for many catalysts. Cesium is a component of highly sensitive photocells, for example, for positioning sensors in satellites (GPS), photomultipliers, and light barriers. Cesium atomic resonance oscillations provide the pulse for the atom clock. The high mobility of the outer electrons are used in scintigrams. Radioactive ^{137}Cs is used in the treatment of tumors.

Ba 56
Barium

Atomic weight
137.3

1808 Sir Humphry Davy (1778–1829). Carl Wilhelm Scheele (1742–1786) discovered the barium ore baryta (barium monoxide), Davy produced the metal.

◀ *The digestive tract shows up very faintly in X-ray scans. The ingestion of barium sulfate allows a good visualization of the details.*

Name: from *barys* (Greek = heavy); barite (or heavy earth) is the weightiest mineral (specific weight = 4.5 g cm^{-3})

Properties
Relatively soft, silver white metal that is attacked by air and water. Slurries of finely ground barite are used as drilling fluid (drilling mud). Insoluble barium sulfate can be swallowed as a contrasting agent without any dangerous side-effects. It also serves as a white pigment in paint, lacquers, and plastics, for example, in power sockets. The soluble salts are toxic. Modern applications include barium titanate (piezoelectric devices), barium ferrite (magnets), and barium cuprate (high-temperature superconductors). All fireworks feature the brilliant green of glowing barium.

La 57
Lanthanum

Atomic weight
138.9

1839 Carl Gustav Mosander (1797–1858) The yttrium ore analyzed by Johan Gadolin (1760–1852) in 1794 and the cerium ore discovered by Jöns Jakob Berzelius (1779–1848) in 1803 were thought to be the same substance until 1839. This mineral was found to contain lanthanum and almost all the lanthanoids.

◀ *Electron microscopes allow a close view of structures. Lanthanum boride provides the electrons.*

Name: from *lanthanein* (Greek = concealed) as it was for a long time hidden with its neighbor cerium

Properties
This soft, silver white metal reacts with air and water. The oxide is applied in optical glasses with high refractive indices (special lenses for powerful cameras and telescopes). Used for special effects in optoelectronics and electronics. Lanthanum exhibits catalytic properties. It is a component of flint and battery electrodes. Lanthanum boride (LaB$_6$) is the superior electron-emitter for electron microscopes. Lanthanum is the first of the series of 14 lanthanides, also called the "rare-earth" metals, whose inner N shells are filled with electrons. They do not belong on the "red list" of endangered species: they are neither rare nor threatened with depletion. China is particularly rich in lanthanide ores.

Ce 58
Cerium

Atomic weight
140.1

1803 Jöns Jakob Berzelius (1779–1848), Wilhelm Hisinger (1766–1852), and Martin Heinrich Klaproth (1743–1817) independently discovered this element.

◀ *Cerium-doped yttrium compounds (aluminates and silicates) cover a very broad spectral range in television tubes.*

Name: Named by M. H. Klaproth after the recently discovered planetoid *Ceres*

Properties
Reactive, gray, relatively soft metal that burns upon heating. Cerium–iron alloys (45–60 % cerium as well as some lanthanum) make up the well-known flintstone in firelighters ("cerium mischmetal"). Cerium is the most common of the so-called "rare earths" and is more abundant than copper and zinc. CeO_2 is a constituent of the ceramic support in catalytic converters. Cerium-containing zeolites play an important role in the production of high-quality petrol. The high temperature resistance is exploited in self-cleaning ovens. Doping with cerium improves activators in phosphors. Cerium is the "fireball" of the metals.

Pr 59
Praseodymium

Atomic weight
140.9

◀ *Protective mask for welders.*

1885 Carl Auer von Welsbach (1858–1929) separated praseodymium and neodymium.

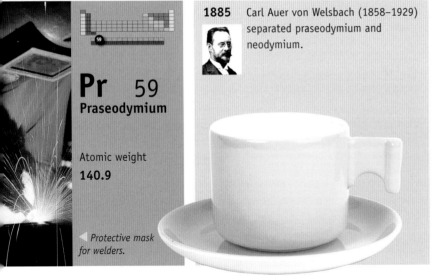

Name: from *prásinos didymos* (Greek = green twin), because it is closely related to neodymium and forms green salts

Properties
Silvery, shiny, ductile metal. The oxide gives a green tint to glasses and glazing, the addition of zirconium oxide results in a striking yellow. Used in protective glass for welding masks as well as in modern sunglasses. The alloy with cobalt and iron is a good permanent magnet. The ceramic supports of catalytic converters are coated with the oxide. Alloys with magnesium are used in airplane engines.

◀ *Praseodymium is rarely encountered in everyday life. However, praseodymium oxide combined with zirconium oxide gives rise to a beautiful yellow glazing, which is very popular.*

Nd 60
Neodymium

Atomic weight
144.2

1885 Carl Auer von Welsbach (1858–1929).

▼ *Neodymium-doped yttrium–aluminum garnet is among the most commonly applied laser material and has broad application (neodymium–YAG).*

Name: from *neos didymos* (Greek = new twin), the newer twin of praseodymium, which it closely resembles

Properties
Silver white metal that tarnishes upon exposure to air and reacts with water. At first was not important. The ruby-colored oxide (Nd_2O_3) tints glass (neophane glass, sunglasses). The real importance of the element only came to light with the advent of laser technology in the "neodymium laser". The famous YAG laser is doped with neodymium. This high-energy laser is ideal for cutting, welding, hardening, engraving, etc. The highlight followed in 1983: $Nd_2Fe_{14}B$ was introduced as the strongest permanent magnet. Its production is simple, and applications boomed immediately! The small magnetic "wunderkind" and its applications are growing and growing (disc drivers, speakers, earphones, model kits (see picture), etc.).

Pm 61
Promethium

Atomic weight
146.9

1945 Jacob A. Marinsky (*1918) as well as L. E. Glendenin and Charles D. Coryll (*1912) detected the element at Oak Ridge. The first conclusive proof was in uranium piles. Uranium fission gives rise to fragments with nuclei of atomic number 61.

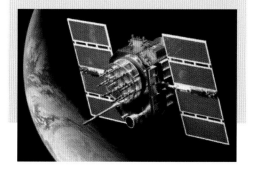

Name: derived from *Prometheus* (Titan from Greek mythology); he brought fire to mankind, for which he was heavily punished

Properties
The only lanthanide of which there is no stable isotope — they all decompose with half-lives between 2.6 and 17.7 years. Strong beta-emitters that are used industrially as thickness gauges. Also suitable as an additive for fluorescent materials. Produced artificially in kg amounts and serves as an energy provider for satellites in radionucleide batteries. Tiny batteries are long-term energy sources for pacemakers.

◀ *Radionuclide batteries provide high-performance satellites with electricity when solar energy is not enough.*

Sm 62
Samarium

Atomic weight
150.4

1879 Paul Emile Lecoq de Boisbaudran (1838–1912)

Name: discovered in the mineral *samarskite* (Samarsky was a Russian geologist)

Properties
Silver white metal that tarnishes in humid air. Led a sheltered life as an additive in glasses and ceramics until the development of samarium–cobalt magnets in the 1970s. Favored for applications in small electrical motors. These iron-free magnets (e.g. $SmCo_5$) are found in miniature motors in watches, toys, etc. Also used in space technology. Without samarium there would have been no Walkman. The magnets are used in nuclear spin tomography. Samarium is indispensable for small microphones and earphones.

◀ *Since the discovery of the magnetic properties of the cobalt alloy, samarium has been the "superstar" of the lanthanides. Music fans have their ears covered with samarium in the form of modern earphones*

Eu 63
Europium

Atomic weight
152.0

1901 Eugène-Anatole Demarcay (1852–1904)

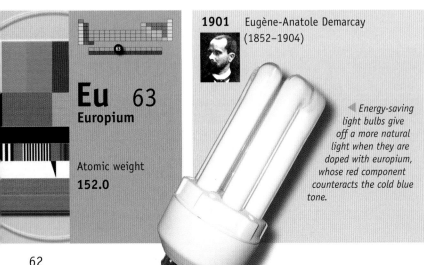

◀ *Energy-saving light bulbs give off a more natural light when they are doped with europium, whose red component counteracts the cold blue tone.*

Name: from the continent *Europe*, where all naturally occurring elements were discovered

Properties
The iron-gray-colored, soft metal is one of the rarest but also the most reactive lanthanide. Its chemistry as well as its properties in superconductors are more the domain of specialists. Nevertheless, it is encountered in everyday applications: the red phosphors in television tubes and computer screens consist of europium-doped yttrium oxisulfide (Y_2O_2S). It is also being used increasingly in energy-saving light bulbs. Europium is the element that can bind the most neutrons (nuclear technology). Europium is rare, but more abundant than gold.

Gd 64
Gadolinium

Atomic weight
157.2

1880 Jean-Charles Gallisard de Marignac (1817–1894) suspected the presence of the element.

1886 Paul Emile Lecoq de Boisbaudran (1838–1912) confirmed and named the new element.

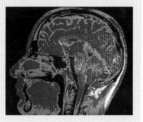

Name: named after the discoverer of the mineral, Johan Gadolin (Finnish chemist)

Properties
Silver white to gray white, shiny, ductile, ferromagnetic metal. Its first important application was in the green phosphors of radar screens (terbium-doped gadolinium oxisulfide). Applied in high-frequency technology (low noise level), in microwave technology, in permanent magnets, and in high-quality glass. Some of its compounds can be injected intravenously as effective contrast agents in NMR tomography (improved diagnosis). Gadolinium has the highest neutron-capture cross-section: the "neutron catcher" for regulation and control in nuclear technology.

◀ *NMR tomography is superior to X-rays for imaging the soft parts of the body. The injection of gadolinium compounds improves the visualization of important details.*

Tb 65
Terbium

Atomic weight
158.9

1843 Carl Gustav Mosander (1797–1858) and Jöns Jakob Berzelius (1779–1848).

Name: from *Ytterby*, the town in Sweden where the corresponding mineral was found

Properties
Silvery metal, that can be cut with a knife. Terbium alloys and additives are widely used in optoelectronics to burn CDs as well as in laser printers. The pronounced magnetostriction (Joule effect) makes "terfenol-D" (terbium–dysprosium–iron) indispensable in sonar technology. The physics of the element appears to be more interesting than its chemistry, in which it is rarely used in catalysis.

◀ *The magnetostriction of terbium alloys has already found many applications. Most recently it was discovered that its compounds can be precipitated on surfaces in highly ordered molecular arrays; this promises interesting effects.*

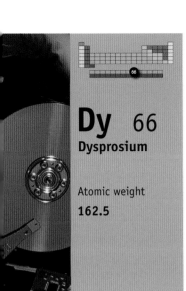

Dy 66
Dysprosium

Atomic weight
162.5

1886 Paul Emile Lecoq de Boisbaudran (1838–1912).

Name: from *dysprósitos* (Greek = difficult to reach or obtain)

Properties
Reactive, hard, silver, shiny metal. Doping with dysprosium increases the coercive force (measure of the strength of a magnet) of neodymium–iron–boron magnets; this allows their use at higher temperatures. The iodide is used in high-performance halogen lamps and intensifies their spectrum in the visible region. The addition of dysprosium increases the sensitivity of dosimeters for gamma rays. Like terbium, it is used in optoelectronics for optical data storage (CDs, DVDs). Dy is used as shielding material in nuclear reactors.

◀ *Presentations rely on strong light sources. High-intensity halogen lamps contain mainly dysprosium iodide.*

Ho 67
Holmium

Atomic weight
164.9

◀ *Strong magnets.*

1878 Per Theodor Cleve (1840–1905); independently Marc Delafontaine (1837–1911) and J.-L. Soret (1827–1890).

Name: from Holmia (Latin = Stockholm)

Properties
Silver-colored, ductile metal that is attacked slowly by air and water. The element exhibits interesting magnetic properties. Found in television tubes. Laser material such as YAG (yttrium–aluminum garnet) doped with holmium (as well as chromium and thulium) can be applied in medicine, especially in sensitive eye operations.

◀ *The technical applications of holmium are limited. In ophthalmology, holmium-doped YAG lasers render a valuable service.*

Er 68
Erbium

Atomic weight
167.3

1843 Carl Gustav Mosander (1797–1858).

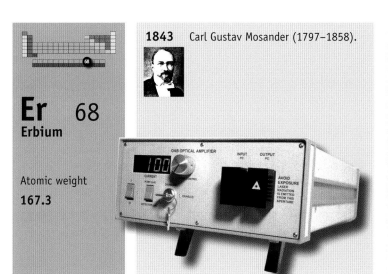

Name: after *Ytterby* in Sweden, where the corresponding ore was found

Properties
Silver gray, ductile metal that was only obtained in pure form in 1934. Bears many similarities to holmium and dysprosium. The pastel pink erbium oxide is used to color glasses. Glasses with erbium oxide act as light amplifiers in the infrared and visible spectral regions. Used in light-conducting glass fibers, which could be of great importance as huge amounts of data could be transferred at the speed of light.

◀ A great future is predicted for glass-fiber technology. Equally essential will be light amplifiers in which the glass fibers are doped with erbium.

Tm 69
Thulium

Atomic weight
168.9

1879 Per Theodor Cleve (1840–1905).

◀ The technical application of thulium is limited. However, the element is becoming increasingly important in special applications of lasers.

Name: from *Thule* (mythical name for Scandinavia)

Properties
The silver gray metal can be cut with a knife, although it only melts at 1545 °C (for comparison, iron: 1538 °C). It is the rarest of the "rare earths", but is nevertheless more abundant than iodine, mercury, and silver. Thulium has few applications, especially because it is relatively expensive. The element occurs naturally as a single isotope, namely ^{169}Tm (compare bismuth). The artificial, radioactive ^{170}Tm is a transportable source of X-rays for testing materials. Occasionally used in laser optics and microwave technology.

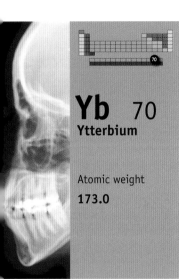

Yb 70
Ytterbium

Atomic weight
173.0

1878 Jean-Charles Galissard de Marignac (1817–1894).
Georges Urbain (1872–1938) and Carl Auer von Welsbach confirmed the existence of the element in 1907.

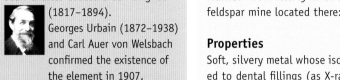

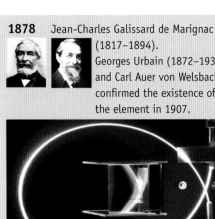

Name: from *Ytterby* in Sweden; four elements were named after the feldspar mine located there: yttrium, erbium, terbium, and ytterbium.

Properties
Soft, silvery metal whose isolation is difficult. Ytterbium fluoride is added to dental fillings (as X-ray contrast agent). Ytterbium is a constituent of rust-free special steels. $Yb_2Co_{13}Fe_3Mn$ holds the magnetic world record, but is too expensive for commercial purposes. The element is occasionally applied in nuclear medicine and radiography. It also activates phosphors that convert infrared rays into visible light.

◀ *Ytterbium-doped glass-fiber lasers show interesting effects.*

Lu 71
Lutetium

Atomic weight
175.0

1907 Georges Urbain (1872–1938). Discovered independently by Carl Auer von Welsbach (1858–1929).

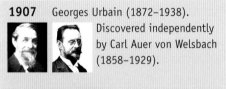

◀ *The production of lutetium is difficult and the demand is low; hence, this metal is about as expensive as gold.*

Name: from *Lutetia* (Latin = Paris)

Properties
The last of the lanthanides, this metal is also the hardest and the densest of them. It is a component of cerium mischmetal. Lutetium has some applications in optoelectronics. Shows great similarities to ytterbium. Its discoverer, Georges Urbain, carried out 15 000 fractional crystallizations to isolate pure lutetium (record!). The element has special catalytic properties (oil industry). ^{176}Lu is generated artificially and is a good beta emitter (research purposes). ^{177}Lu has a half-life of six days and is used in nuclear medicine.

Hf 72
Hafnium

Atomic weight
178.5

1923 George de Hevesy (1885–1966; Nobel Prize for chemistry 1943) and Dirk Coster (1889–1950). The search for this element was long, and it was eventually found as a companion of zirconium minerals by means of X-ray spectroscopy.

◀ *Hafnium is seldom encountered in everyday life. Alloys with hafnium are used in flash devices.*

Name: from *Hafnia* (Latin = Copenhagen), as both researchers correctly interpreted their results at the institute of Niels Bohr.

Properties
The silvery, shiny, ductile metal is passivated with an oxide layer. Chemically very similar to and always found with zirconium (like chemical twins, with almost identical ionic radii); the two are difficult to separate. Used in control rods in nuclear reactors (e.g. in nuclear submarines), as it absorbs electrons more effectively than any other element. Also used in special lamps and flash devices. Alloys with niobium and tantalum are used in the construction of chemical plants. Hafnium dioxide is a better insulator than SiO_2. Hafnium carbide (HfC) has the highest melting point of all solid substances (3890 °C; record!).

▶ *Nuclear submarine*

Ta 73
Tantalum

Atomic weight
180.9

1802 Anders Gustaf Ekeberg (1767–1813) Heinrich Rose (1795–1864) distinguished tantalum from niobium.

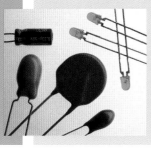

◀ *High-performance capacitors are indispensable in electronics (computers, televisions, cars,...). About 20 billion are produced each year.*

Name: derived from *Tantalos* (from Greek mythology: father of Niobe; although he stood in water, he could not quench his persistent thirst, a punishment from the gods)

Properties
Shiny, silvery metal. Soft even though it only melts at 2996 °C. It is 50 % heavier than lead. Its isolation was highly problematic. The element exhibits high corrosion resistance (chemical apparatus, surgical instruments, jet nozzles). Tantalum oxide has a high dielectric constant and is therefore ideal for high-performance capacitors, which are found in almost all electronic devices. Tantalum carbide is used in metal processing. Suitable for implants (e.g. knee pins), as it is completely biologically inert.

W 74 Tungsten

Atomic weight **183.8**

1783 Fausto de Elhuyar (1755–1833) and his brother Juan José prepared it by the reduction of WO₃ with carbon.

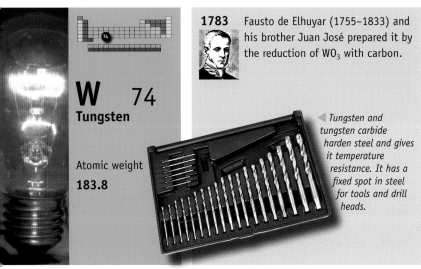

Tungsten and tungsten carbide harden steel and gives it temperature resistance. It has a fixed spot in steel for tools and drill heads.

Name: from *tung sten* (Swedish = heavy stone); chemical symbol: wolfram derived from *Wolfrahm* (Middle High German): like a wolf, it stole the desired tin in melting furnace

Properties
Silvery, shiny metal. Has the highest melting point (3410 °C; record!) and the second-highest boiling point (5555 °C) of all the metals. Best suited for cathodes in X-ray tubes. Everybody knows the glowing filament as well as the legendary hard steel. Tungsten carbide objects are even harder. Alloys with cobalt are called widia (from German: "wie Diamant" = like diamond). Ideally suited for drill bits and cutting tools as well as jet nozzles in rockets. The element exhibits good catalytic properties. Also used in the production of semiconductors. There is a series of wolframate pigments. Remarkable element: hard and versatile.

Re 75 Rhenium

Atomic weight **186.2**

1925 Walter Noddack (1893–1960), Ida Tacke (1896–1978), and Otto Berg (1874–1939).

Discovered by means of X-ray spectroscopy in columbite after a systematic search.

Name: from *rhenus* (Latin = Rhine)

Properties
Very rare, silver white metal that hid itself behind molybdenum for a long time. Has the highest boiling point of all elements (5597 °C; record!) as well as the second-highest melting point (3180 °C). Is heavier (and more expensive) than gold. Ideally suited for use in heating elements in ovens. Valuable alloy component for heat-resistant materials. Has recently been increasingly applied as a catalyst, for example, as a "reform catalyst" in the production of petrol. In thermocouples, the metal allows temperatures as high as over 2000 °C to be measured.

◁ *The heavily stressed anode in high-performance X-ray tubes is coated with a rhenium–tungsten alloy.*

Os 76
Osmium

Atomic weight
190.2

1803 Smithson Tennant (1761–1815).

There are no pictures, only letters (see iridium).

Found as an admixture in crude platinum, could not be dissolved in aqua regia.

◀ *Fritz Haber first used osmium for the synthesis of ammonia. The rare and expensive element was soon replaced by the cheapest: iron.*

Name: from *osme* (Greek = smell)

Properties

Silvery, shiny, and hard. Unique metal, gives off an odor as it forms volatile OsO_4 on the surface (oxidation states: 8!). Osmium is the densest element (22.6 g cm^{-3}; record!). Was replaced in filaments (Osram) by the cheaper tungsten. Used in platinum alloys and as a catalyst. Haber's first catalyst in ammonia synthesis was osmium, which fortunately could be replaced by doped iron. The addition of as little as 1 to 2 % of this expensive metal increases the strength of steel (e.g. fountain-pen tips, early gramophone needles, syringe needles).

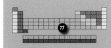

Ir 77
Iridium

Atomic weight
192.2

1803 Smithson Tennant (1761–1815).

Fell from his horse to his death during his trip through Europe.

Found as an admixture in crude platinum, could not be dissolved in aqua regia.

Name: from *iris* (Latin = rainbow), owing to its brilliant salts

Properties

Very hard, steel-gray metal. Hardens platinum. The International Prototype Meter in Paris consists of a Pt–Ir alloy. Its hardness and corrosion resistance is exploited in fountain-pen tips, spark plugs in powerful engines (airplanes), and electrical contacts. Used as a material in shells for nuclide batteries in satellites. Responsible for the iridescent properties of vapor-treated sunglasses.

◀ *Vapor deposition allows the production of sunglass lenses with an extremely thin protective coating of iridium.*

Pt 78
Platinum

Atomic weight
195.1

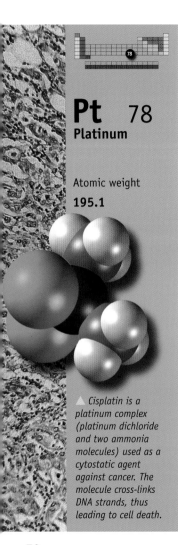

▲ Cisplatin is a platinum complex (platinum dichloride and two ammonia molecules) used as a cytostatic agent against cancer. The molecule cross-links DNA strands, thus leading to cell death.

Platinum was unknown in Europe in antiquity.

1735 The Spanish natural scientist Antonio de Ulloa (1716–1795) brought metal samples from South America.
William Brownrigg (1711–1800) gave the first scientific description, which was revealed in 1750.

▼ *Platinum is a true treasure in catalysis. Many reactions proceed faster and more effectively in the presence of platinum; this is specially true for oxidations.*

Name: from *platina* (Spanish = little silver)

Properties
Very shiny metal that is ductile in its pure form. Extremely resistant, dissolves only in hot aqua regia. At first despised because it interfered with the extraction of gold, until Döbereiner and Davy recognized its catalytic properties. In this field, no element is as versatile as platinum: oxidations (e.g. $NH_3 \rightarrow HNO_3$ as well as $CO \rightarrow CO_2$ in automobile catalytic converters) are the best-known examples. It also catalyzes hydrogenations. Its high resistance is also commonly exploited: platinum crucibles, standard meter, standard kilogram. Cisplatin complexes are very effective anticancer agents. The element is about twice as expensive as gold and also used in jewelry. Platinum crucibles decorate every lab.

◀ *High chemical resistance makes platinum (with about 10 % iridium) an ideal material for the standard meter and the standard kilogram. Furthermore, platinum is about twice as dense as lead.*

◀ *Platinum crucibles and apparatus are the "silverware" of any good laboratory. Reactions under aggressive conditions can often only proceed in platinum vessels.*

Au 79
Gold

Atomic weight
197.0

As it occurs naturally in its free form, gold has been known since the most ancient cultures and used for jewelry and cult objects. The alchemists assigned gold to the sun.

▲ Virgin gold in golden veins is every gold digger's dream. Mostly, river-worn nuggets had to be tediously sorted from alluvial sand.

▶ Its scarcity and imperishable shine have fascinated man since prehistoric times. As soon as money was invented, gold coins had the highest value. This is still true today.

Name: from *ghel* or *ghol* (Indo-Germanic = yellow, shiny); chemical symbol: *aurum* (Latin = gold)

Properties
Gold is the most ductile of all elements (gold sheet). Can be worked so thin that light goes through; it then looks green. Gold is among the rarest metals. The Earth's crust contains 3 to 5 g per ton. Since the early ages, about 120 000 tons have been extracted (cube with 18-m edges). Gold is inert, hence also very durable and permanently shiny. Earlier the basis for currencies, today it is used in dental care, electronics (platinum circuits), and for corrosion protection in cable plugs. Gold is the underhanded King of the Periodic Table in whose name much mischief has been done!

▶ Of the approximately 3000 tons mined annually, only a tiny portion ends up in technical applications. Circuit boards and plug connection points are protected against corrosion by gold. This is where gold is really valuable.

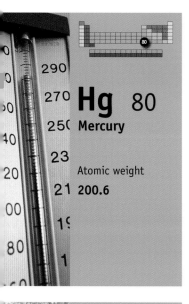

Hg 80
Mercury

Atomic weight
200.6

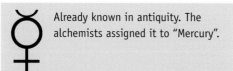

Already known in antiquity. The alchemists assigned it to "Mercury".

Name: from *hydragyrum* (Latin = liquid silver); *mercury* (English, from the name of the planet); *quecsilver* (Old High German = live silver)

Properties
Mercury is the only liquid metal (melting point: –39 °C). The bright red sulfide (cinnabar used as a pigment) very quickly drew attention to this element and hence its early discovery. Used in thermometers, manometers (blood pressure is still measured in mmHg), electrical equipment (toggle switches), batteries, chlorine-alkaline electrolysis, and was previously used very often as an amalgam (mercury alloy), especially in fillings (in this case, silver). When the high toxicity of methyl silver was recognized, its application was drastically cut down. Today still found in barometers, fluorescent tubes, mercury-diffusion pumps (high-vacuum) as well as in the opposite sense in high-pressure lamps (e.g. street lighting).

◀ *Mercury drops were often encountered when a thermometer was broken. Mercury is poisonous and banned from households.*

Tl 81
Thallium

Atomic weight
204.4

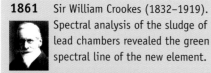

1861 Sir William Crookes (1832–1919). Spectral analysis of the sludge of lead chambers revealed the green spectral line of the new element.

Name: : from *thallos* (Greek = green twig or shoot); gives flames an intense green color

Properties
Soft, pale gray metal with a green spectral line. Used in green signal rockets. Thallium silicate glasses have a high refractive index and are very transparent to infrared light. Used especially in infrared detectors for copiers and fax machines. The element is highly toxic, as thallium ions displace potassium in essential life functions (the odorless and tasteless sulfate was considered a good rat poison). Murders by thallium poisoning could be proved by spectral analysis, even after cremation of the body. Tiny traces of the radioactive isotope can be useful in medical diagnostics. Alloys with mercury are used in low-temperature thermometers.

◀ *Thallium still serves the environment; however, not as rat poison, but in infrared cameras to detect heat loss.*

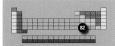

Pb 82
Lead

Atomic weight
207.2

Already known in antiquity, assigned by the alchemists to Saturn. They tried to convert lead into gold.

▲ Galenite (PbS$_2$) in mineral veins is very conspicuous, which led to the early discovery of the metal.

Name: symbol from *plumbium* (Latin = lead)

Properties
Soft, "lead-gray" metal that tarnishes upon exposure to air and is therefore resistant to corrosion. It is among the veterans of the elements. Early application as water pipes. The Romans allegedly fortified wine in lead vessels (chronic lead poisoning?). The discovery of guns started its dubious career as lead shot. Then used in car batteries and as a petrol additive. As lead is toxic (binds to SH groups of proteins), its applications have been significantly reduced. Batteries should be recycled. Lead might be past the peak of its use, but it is irreplaceable for radiation shields (X- and gamma rays). Lead-crystal glasses are still highly valued for their shine. Since 2003, the lead-208 isotope is considered the heaviest stable element (record!).

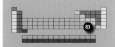

Bi 83
Bismuth

Atomic weight
209.0

Mentioned by Basilius Valentinus and Paracelsus, comprehensively described by Georgius Agricola (1495–1555) in his book "De re metallica" (1556).

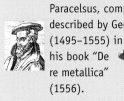

▶ The low melting point of bismuth and its alloys is exploited in fire sprinklers.

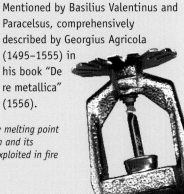

Name: originally *wismut* (origin not certain, perhaps "extracted from the fields" or "white mass")

Properties
The brittle, silvery, shiny metal was long considered the last stable element of the Periodic Table. In 2003 it was unmasked as an extremely weak alpha emitter (half-life: 20 billion years). Like thulium, there is only one isotope. Bismuth alloys have low melting points (fuses, fire sprinklers). As an additive in tiny amounts, it imparts special properties on a range of metals. Applied in electronics and optoelectronics. The oxichloride (BiOCl) gives rise to pearlescent pigments (cosmetics). As bismuth is practically nontoxic, its compounds have medical applications. The basic oxide neutralizes stomach acids. A multitalented element. Crystallizes with an impressive layering effect (see right).

Po 84
Polonium

Atomic weight
209.0

1898 Marie Curie (née Sklodowska; 1867–1934; Nobel Prize for physics 1903 and for chemistry 1911) detected it in pitchblende after laborious enrichments.

◀ *Polonium was considered a rarity. With beryllium, the alpha emitter generates a strong neutron flux. Used as the detonator in the first atomic bombs.*

Name: from *Poland* (home of Marie Curie).

Properties

Radioactive, silver gray metal. Product of uranium decay. Such a strong alpha emitter that it heats itself and in the dark creates a light blue glow in the surrounding air through excitation. Enduring energy source in satellites. Serves as a source of alpha rays in research. In combination with beryllium, makes an effective neutron source: "detonator" for nuclear weapons. This would certainly not have been to the liking of the idealistic Marie Curie, who discovered radium and polonium in her scientific research and described them in her dissertation. In 1903, together with her husband Pierre and with A. H. Becquerel, she was awarded the Nobel Prize for physics.

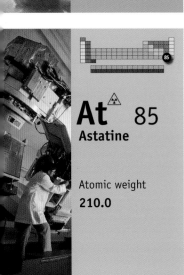

At 85
Astatine

Atomic weight
210.0

1940 Emilio Gino Segrè (1905–1989; Nobel Prize for physics 1959), together with Dale Raymond Corson (*1914) and Kenneth Ross Mackenzie (*1912) obtained the element in tiny amounts by bombardment of bismuth with alpha particles. Halogen with no stable isotope.

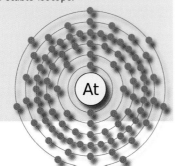

Name: from *astatos* (Greek = unstable)

Properties

The element exists as an intermediate in uranium and thorium minerals through their decay. There is no stable isotope. The longest-living isotope has a half-life of 8.3 hours. In the crust of the Earth, the total steady-state mass is estimated at a few tens of grams. Thus astatine is the rarest element (record!). A few atoms of this relative of iodine can be found in all uranium ore. It exhibits certain metallic properties.

◀ *Astatine is isolated in tiny amounts from reactor materials. The Bohr atomic model shows the tightly packed electron shell. One can formally "see" the instability. It was the last of the 92 naturally occurring elements to be found.*

Rn 86
Radon

Atomic weight
222.0

1900 Friedrich Ernst Dorn (1848–1916). In 1900 Ernest Rutherford (1871–1937) detected the radon-220 isotope as a decay product of thorium. In the same year, Dorn showed the radon-222 isotope to be a decay product of radium.

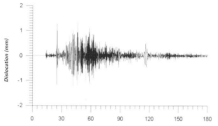

Name: derived from *radium*

Properties
This short-lived radioactive gas arises from the decay of uranium or thorium. It is the rarest noble gas and is of no technical importance. It seeps from the floors of rooms, especially in the case of granite subsoil. The danger is low, but airing is nevertheless recommended. Although an unwelcome guest in homes, in sharp contrast it is believed to have relieving properties in mineral springs. An increase in radon emissions signals danger in earthquake areas, where the levels are closely monitored. Large amounts are given off during volcanic eruptions.

◁ *An increased emission of radon is a reliable warning sign of an impending earthquake.*

▷ *During a volcanic eruption, a lot of ash is blasted into the atmosphere. The gases, such as sulfur dioxide and radon, remain invisible.*

Fr 87
Francium

Atomic weight
223.0

1939 Marguerite Perey (1910–1975). The element "eka-cesium" had long been suspected. Was detected as a short-lived intermediate product in the decay series of actinium.

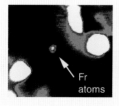

◁ *Francium is extremely unstable. High-tech and much luck are necessary to detect these atoms.*

Name: from *France*, the country of its discovery

Properties
Strongly radioactive, short-lived element that can be found in tiny amounts in uranium ores. It arises fleetingly from ^{235}U in its decay chain through actinium (^{227}Ac). Is only of scientific value as it has a maximum half-life of about 22 minutes. Nevertheless, in its short existence it is the atom with the largest diameter (0.270 nm; cesium: 0.265 nm).

▷ *As francium arises from the decay series of ^{235}U, every piece of uranium ore contains at least a few atoms.*

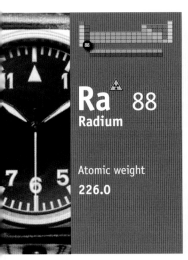

Ra 88
Radium

Atomic weight
226.0

1898 Marie Curie (1867–1934) and Pierre Curie (1859–1906); Nobel Prize for physics 1903. Only 0.1 g of radium chloride was isolated from about 500 kg or uranium pitchblende (1:5 million).

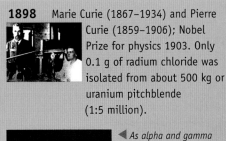

◀ *As alpha and gamma emitters, radium isotopes allow other material such as zinc sulfide and glass to glow in the dark. Its applications have been cut back owing to the dangers involved in its production.*

Name: from *radius* (Latin = ray)

Properties
The radioactive element is a silvery, shiny, soft metal that is chemically similar to calcium and barium. It is found in tiny amounts in uranium ores. Its radioactivity is a million times stronger that that of uranium. Famous history of discovery (in a shed). Initially used in cancer therapy. Fatal side effects. Small amounts are used in luminous dyes. Radium was of utmost importance for research into the atom. Today its reputation is rather shaky as its decay gives rise to the unpleasant radon (see earlier). In nuclear reactors, tiny amounts of actinium are formed from radium.

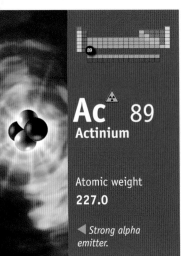

Ac 89
Actinium

Atomic weight
227.0

◀ *Strong alpha emitter.*

1899 André Louis Debierne (1874–1949)

1902 Friedrich Oskar Giesel (1852–1927). Detected independently by the two researchers in uranium pitchblende.

Name: from *aktinos* (Greek = ray)

Properties
Soft, silver white metal that is isolated in the tiniest of amounts. All isotopes are radioactive, the longest-lived has a half-life of 22 years. The element is an intermediate in the decay series of ^{235}U. Strong alpha emitter that is used in radioactivation analysis and forms an effective neutron source with beryllium.

◀ *Actinium occurs naturally in such small amounts that isolation is impossible. Considerable amounts can be isolated from reactor material.*

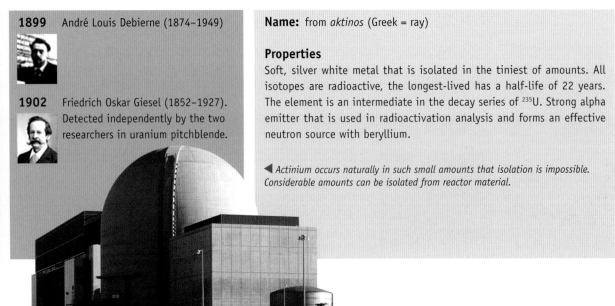

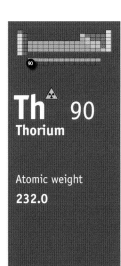

Th 90
Thorium

Atomic weight
232.0

1829 Jöns Jakob Berzelius (1779–1848) discovered ThO$_2$ in 1828. The element itself (in impure form) was produced by heating potassium with potassium thorium fluoride.

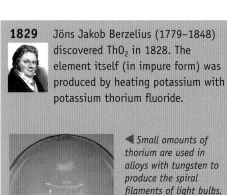

◀ Small amounts of thorium are used in alloys with tungsten to produce the spiral filaments of light bulbs. Higher temperature generate a brighter light.

Name: from *Thor* (god of thunder in Nordic mythology)

Properties
Radioactive, silvery, soft metal that is rare, but nevertheless is three times more abundant than uranium. Thorium is a weak alpha emitter (half-life: 14 billion years). The metal is used to prepare alloys with high heat resistance (e.g. turbine blades). Thorium dioxide is extremely heat-resistant and has the highest melting point (3300 °C) of all the oxides. This fire-resistant material earlier enjoyed large-scale use in incandescent gas mantles in the household and in street lighting. Until the 1940s colloidal ThO$_2$ was injected as an effective contrast agent for X-rays. As it is practically not excreted, dangerous after-effects appeared later. Thorium oxide or carbide serves as breeder material in high-temperature reactors. This is significant for countries with little access to uranium. One such reactor is operational in India and does not give rise to plutonium.

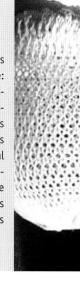

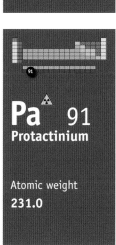

Pa 91
Protactinium

Atomic weight
231.0

1913 Kasimir Fajans (1887–1975) and O. H. Göhring detect a short-lived isotope.

1917 Otto Hahn (1879–1968; Nobel Prize for chemistry 1944) and Lise Meitner (1878–1968) as well as Frederik Soddy (1877–1956; Nobel Prize for chemistry 1921) discovered a further isotope in uranium pitchblende.

1927 Produced in a pure form by Aristid Victor Grosse (1905–1985).

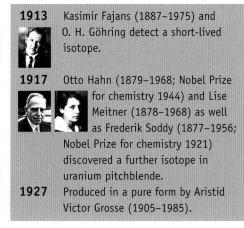

Name: from *protos* (Greek = first); i.e., before actinium

Properties
Radioactive, silvery metal of which only about 125 g exists worldwide, isolated from reactor material. Protactinium occurs in the decay series of ^{238}U (K. Fajans) as ^{234}Pa. It also occurs in that of ^{235}U; this isotope, ^{231}Pa, was discovered by L. Meitner and O. Hahn. The element is only of scientific interest.

▶ In every piece of uranium ore, such as pitchblende, there are 0.1–0.3 ppm of protactinium as an intermediate in the decay series. The isolation is not

U 92
Uranium
Atomic weight **238.0**

1789 Martin Heinrich Klaproth (1743–1817) discovered the new metal in uranium pitchblende.

1841 Eugène M. Peligot (1811–1890) first isolated the metal.

▶ The fascinating uranium glasses were already known throughout empires before its discovery. They then became sought-after articles.

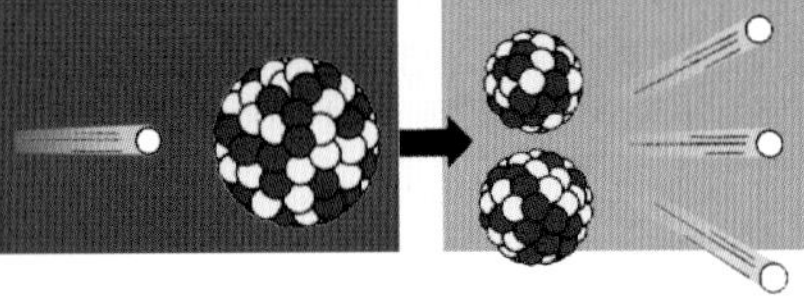

Name: named after the planet *Uranus*, which was discovered as a new planet shortly before (1781); when O. Hahn split the uranium atom in 1939, we entered the atomic age

Properties
Radioactive, silver gray metal that consists of the isotopes ^{238}U (99.3 %) and ^{235}U (0.7 %). The latter is split by slow neutrons to give rise to large amounts of energy and the release of two or three neutrons, which initiate a chain reaction. Nuclear reactors provide emission-free energy. The problem is the disposal of nuclear waste. The enrichment of ^{235}U with gaseous UF_6 was first successful during the construction of the atomic bomb. This tragedy marks a low in human history and still today is a major argument against nuclear energy for peaceful purposes. Depleted ^{238}U has a half-life of 4.47 billion years and is practically not radioactive. It is as dense as gold. Used in munitions and trim weights in aircraft. Uranium gives glass a characteristic shimmering green color. The extremely weak alpha radiation is not dangerous.

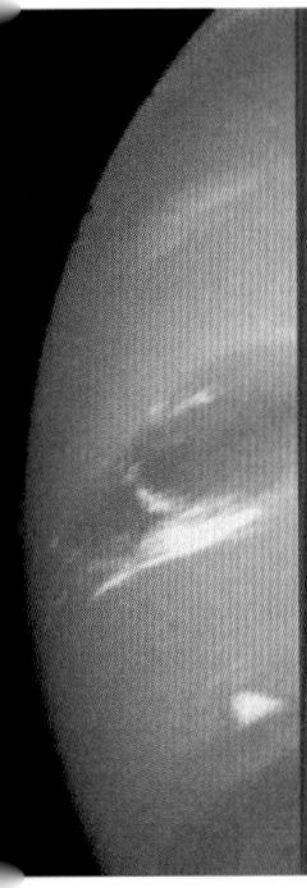

Np 93
Neptunium
Atomic weight **237.0**

1940 Edwin Mattison McMillan (1907–1991; Nobel Prize for chemistry 1951) and Philip Hauge Abelson (1913–2004).

Identification of the isotope ^{239}Np, which is generated by slow-neutron bombardment of ^{238}U and subsequent beta decay.

α-decay
β⁻-decay

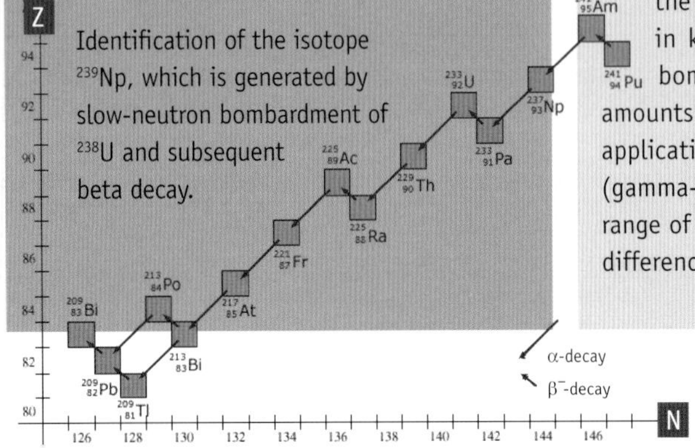

Name: named after *Neptune* (the planet after Uranus)

Properties
Radioactive, silver-colored, highly reactive metal. Marks the start of the series of the artificial transuranium elements. Can be extracted in kilogram amounts from burnt out reactor fuel. Arises from the bombardment of uranium with neutrons. Naturally formed in tiny amounts from uranium by neutron capture and subsequent beta decay: application as neutron detector as well as in Mössbauer spectroscopy (gamma-ray resonance spectroscopy). Neptunium has the biggest liquid range of all the elements: melting point: 630 °C, boiling point: 3900 °C, difference: 3270 °C (cf. gallium).

◀ The picture shows one of the three natural decay series according to which heavy, radioactive nuclei eventually decay to stable lead atoms.

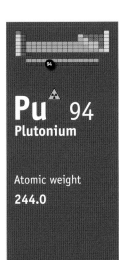

Pu 94
Plutonium

Atomic weight
244.0

1940 Glenn Theodore Seaborg (1912–1999; Nobel Prize for chemistry 1951), Edwin Mattison McMillan (1907–1991; Nobel Prize for chemistry 1951), together with Arthur C. Wahl and Joseph W. Kennedy. Bombardment of ^{238}U with cyclotron-accelerated deuterons gave rise to the isotope ^{238}Pu after some intermediates. Plutonium was the first element to be generated artificially in large amounts.

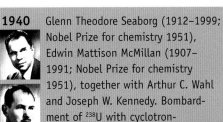

▶ *Fuel rods.*

Name: named after *Pluto*, which recently lost its status as a planet; Greek god of the underworld and wealth; G. T. Seaborg recognized the analogy between the lanthanoids and actinoids

Properties
Silvery metal that chemically resembles samarium. Generated by neutron bombardment of ^{238}U in nuclear reactors. This so-called "breeding" allows uranium to be used completely to produce energy. Strong alpha emitter; the radiation has a short range. The metal can be handled without any problems. However, if dust particles are inhaled, they destroy the surrounding cells. Discovered in a cyclotron in Berkeley upon bombardment of uranium with deuterons. When the fission potential was recognized, production began within the frame of the Manhattan Project. The second atomic bomb (Nagasaki) was a plutonium bomb. As a nuclear fuel and a warfare agent, plutonium is the best-known but also the most controversial of the transuranium elements. Applied as an energy source in satellites and space stations.

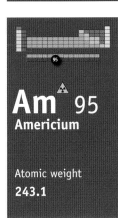

Am 95
Americium

Atomic weight
243.1

1944 Glenn Theodore Seaborg (1912–1999) and Albert Ghiorso (*1915) together with Ralph A. James and Leon Owen Morgan (1919–2002). The difficult separation from the neighboring elements was achieved by ion exchange.

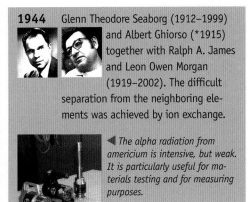

◀ *The alpha radiation from americium is intensive, but weak. It is particularly useful for materials testing and for measuring purposes.*

Name: named after the continent where it was discovered, in Chicago in 1944 in the world's first nuclear reactor

Properties
Light-silver-colored element generated from a plutonium isotope (^{241}Pu) by beta decay. Never detected in nature. Chemically similar to Europium. A few tons have been produced throughout the world through regeneration of fuel rods. Americium is a good source of alpha rays. Hence it is suitable to measure thicknesses, as a detector in smoke alarms, and for the activation analysis of the tiniest amounts of substances.

▶ *Industrial smoke detector on the basis of americium.*

Cm 96
Curium
Atomic weight
247.1

Bk 97
Berkelium
Atomic weight
247.1

Cf 98
Californium
Atomic weight
251.1

1944 Glenn Theodore Seaborg (1912–1999), together with Ralph A. James and Albert Ghiorso (*1915). Irradiation of a plutonium-239 sample with alpha particles and enrichment of ^{242}Cu by chemical methods.

Name: in honor of *Marie Curie* (1867–1934) and *Pierre Curie* (1958–1906)

Properties
Silvery, artificial element generated by beta decay from a plutonium isotope (^{239}Pu). Chemically similar to gadolinium. Like Eu and Gd, Am and Cm are difficult to separate. It can be produced in kilogram amounts. The most common isotope is ^{244}Cm with a half-life of 18.1 years. Is used for thermoelectric nuclide batteries in satellites and pacemakers. It is strongly radioactive and hence also suitable for material analysis.

1949 Glenn Theodore Seaborg (1912–1999), together with Stanley Gerald Thompson (1912–1967) and Albert Ghiorso (*1915). The bombardment of americium-241 with alpha particles led to element 97 with atomic mass number 243. The enrichment involved chemical methods, as the properties of the element were assumed to be analogous to those of the lanthanides.

Name: named after its place of discovery, *Berkeley*, where Ernest O. Lawrence built the first cyclotron and where most artificial elements were discovered

Properties
Silver-colored, artificial element, first produced by bombarding americium with helium ions. Chemically similar to terbium. Bk begins the series of elements that are generated by bombardment of artificial elements with helium ions or neutrons. To date produced in mg amounts. Only of scientific interest.

1950 Glenn Theodore Seaborg (1912–1999), Albert Ghiorso (*1915), together with Stanley Gerald Thompson (1912–1967) and Kenneth Street (1920–2006). Generated by bombardment of ^{242}Cm with alpha particles.

Name: named after *California*, where Cf was first produced at the University of Berkeley by bombardment of Curium with alpha particles.

Properties
Silvery white, artificial element that is also generated by intensive bombardment of plutonium with neutrons. It is a strong ("hot") neutron emitter and is used in microgram quantities in nuclear medicine. This reliable neutron source is also used in industry and science (for activation analysis).

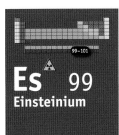

Es 99
Einsteinium
Atomic weight
252.1

1952 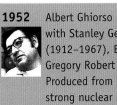 Albert Ghiorso (*1915), together with Stanley Gerald Thompson (1912–1967), Bernard G. Harvey, and Gregory Robert Choppin (*1927). Produced from plutonium-239 by strong nuclear bombardment.

Name: in honor of Albert *Einstein* (1879–1955)

Properties
Unstable, silvery metal. The element was first discovered in the fallout from the first hydrogen bomb on the Bikini Atoll (1952), later produced by neutron bombardment of plutonium. Half-lives of the isotopes range from 20 to 401 days. "Relatively short-lived" in comparison to Einstein's formula $E = m \cdot c^2$, which is valid forever. Only of scientific interest.

Fm 100
Fermium
Atomic weight
257.1

1952 Albert Ghiorso (*1915),), together with Gregory Robert Choppin (*1927), Stanley Gerald Thompson (1912–1967), and Bernard G. Harvey. Produced from plutonium-239 by strong nuclear bombardment.

Name: in honor of Enrico *Fermi* (1901–1954), who built the first nuclear reactor in Chicago

Properties
Like einsteinium, this unstable element was discovered in the fallout from the first hydrogen bomb. To date, only fragments in microgram amounts can be isolated. ^{258}Fm ends the series of transuranium elements that can be produced in a reactor by neutron bombardment. The longest-lived isotope decays with a half-life of 100 days

Md 101
Mendelevium
Atomic weight
258.1

1955 Glenn Theodore Seaborg (1912–1999), Albert Ghiorso (*1915), together with Bernard G. Harvey, Gregory Robert Choppin (*1927), and Stanley Gerald Thompson (1912–1967). Bombardment of einsteinium-253 with alpha particles allowed the detection of 17 atoms of element 101.

Name: in honor of Dimitri Ivanovich *Mendeleev* (1834–1907)

Properties
Radioactive, short-lived element. The longest-lived isotope (^{256}Md) has a half-life of 55 days. To date, only a few atoms have been prepared by a nuclear reaction between einsteinium and helium nuclei in a particle accelerator.

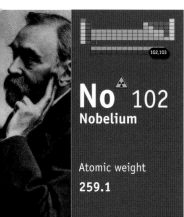

No 102
Nobelium

Atomic weight
259.1

1958 Albert Ghiorso (*1915) and Glenn Theodore Seaborg (1912–1999) together with Torbjörn Sikkeland and J. R. Walton could ambiguously prove the presence of the short-lived element, which had been suspected in 1957 at the Nobel Institute in Stockholm.

Name: in honor of Alfred *Nobel* (1833–1896), inventor of dynamite and founder of the Nobel Prize

Properties
The name was debated. The priority for the production lay with the Institute for Nuclear Research in Dubna (Russia), under the leadership of G. N. Flerov, where ^{238}U was bombarded with ^{22}Ne. The half-life was about 2.7 seconds. At the University of California, bombardment of curium with carbon gave rise to an isotope with a half-life of 58 minutes. The IUPAC commission suggested the name "lerovium", but nobelium persisted.

Lr 103
Lawrencium

Atomic weight
262.1

1961 Albert Ghiorso (*1915) together with Torbjörn Sikkeland, Almon Larsh, and Robert M. Latimer obtained the element by bombarding californium with boron ions.

Name: in honor or Ernest O. *Lawrence* (1901–1958; Nobel Prize for physics 1939), inventor of the cyclotron

Properties
The element was generated by bombardment of californium with boron in a linear accelerator. The priority is debated. Isotopes of the elements were observed both by the group of Glenn T. Seaborg and by that of G. N. Flerov in Dubna. IUPAC proposed that the priority be shared. The longest-lived isotope has a half-life of 200 minutes. Lawrencium ends the series of actinides, as the 5f level is fully occupied with 14 electrons.

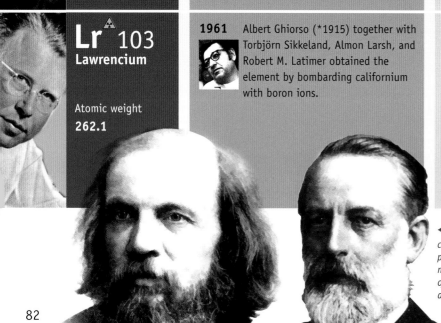

◀ The detailed knowledge and vision of Dimitri Ivanovich Mendeleev as well as the precise objectivity of Lothar Julius Meyer led to the recognition of periodicity as the ordering principle of the elements. They laid the foundation for our knowledge today of what makes up the world and the universe. This is one of the greatest cultural achievements of mankind. They showed impressively what humans are capable of through their insight and thirst for knowledge.

Where to from Here?

Our description of the Periodic Table ends with the element lawrencium with atomic number 103. Its discovery in 1961 ended the actinide series. But what comes next? In 1949, a shell model that resembled the Bohr model for electrons was proposed also for the atomic nucleus. According to this model, closed nucleon shells should give rise to stable nuclei. This theory predicted so-called magic numbers for particularly stable atoms with atomic numbers (number of protons) 2, 8, 20, 28, 50, 82, 126, and 184. Helium (2), oxygen (8), calcium (20), and nickel (28) confirmed the theory. Also tin (atomic number 50) is conspicuous as the element with the largest number of stable isotopes. And lead with its 82 protons is the heaviest nonradioactive element. In elements with higher atomic numbers, the repulsive forces of the protons gain the upper hand and can no longer be compensated for by the number of neutrons. The larger the nucleus, the greater the excess of neutrons required to achieve stability.

There are seven naturally occurring radioactive elements; from uranium, the elements are all unstable and were prepared artificially. This is an astonishing phenomenon. Not only did evolution produce a being that was in the position to discover the Periodic Table, it gave it the ability to produce elements that do not exist in nature. Needless to say, we shall no go any deeper into this aspect of natural philosophy.

Back to the facts. The use of accelerators as fusion reactors first in 1940 in Berkeley (USA), later in Dubna (Russia), and then in Darmstadt (Gesellschaft für Schwerionenforschung; Institute for Heavy-Ion Research) allowed the expansion of the series of elements up to atomic number 116. This means that 24 artificial elements after uranium have been produced and identified. In most cases, the half-lives are extremely short and the few atoms could only be identified from their decay products. The International Union of Pure and Applied Chemistry (IUPAC) suggested that the ultra-heavy elements be given the Latin names of their atomic numbers. Further research would hardly be justified, were it not for the nuclear model that predicts a particularly stable nucleus at atomic number 126. Is there really such a "Stability Island"? The great mathematician David Hilbert (1862–1943) had the following epitaph engraved on his gravestone: "We must know, we will know." Let's wait and see.

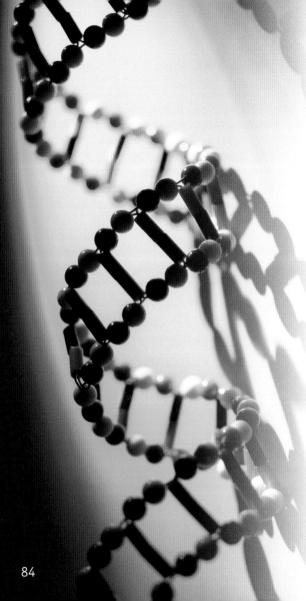

The Elements and Life

We know what the elements are. But what is life? From the point of view of natural science, living systems must fulfill three criteria:

1. **Metabolism (maintains the viability of the individual)**
2. **Reaction to foreign stimuli**
3. **Reproduction (survival of the species)**

These criteria are plausible and can be investigated by scientific methods. Nevertheless, a phenomenon is missing that is intrinsic to all life forms:

4. **All life wants to live!**

Why this is so cannot be explained. Equally inexplicable is the trend towards the more-complex structures and systems observed in evolution. Even if we conclude from this that life in its beginning had to be relatively simple, it was still not primitive. Manfred Eigen (Nobel Prize 1967) in particular pointed out that one property of matter must have played a decisive role: the ability to self-assemble. From there he postulated that "Life comes into existence when the conditions for it are suitable". Could a different life have come into existence had the conditions been different? We don't know. Instead, we know a lot about life that we can observe and investigate.

Let us begin with the simple question as to how many elements were involved in the adventure of life. The central position is occupied by carbon with its versatile reactivity profile. In terms of amount, it is exceeded by hydrogen and oxygen. This is because life most probably began in water and all life forms contain a lot of water, for example, humans with about 60 %. The record at 99 % is held by the jellyfish. Nitrogen is the key element of proteins. From here, the amounts clearly decrease. In total, the function of 17 main-group elements (old nomenclature) and 10 transition-metal ions (i.e. 27 elements) has been elucidated. Tungsten occurs very rarely, yet has been found in some organisms.

The so-called trace elements in general have a catalytic function or are involved in regulatory processes. Some elements exhibit contradicting properties. For example, selenium in large amounts is highly toxic, but a deficiency can have serious health repercussions. It all comes down to the dosage, but this we have known, at the latest, since Parcelsus.

Which elements do life forms require to code their genetic inheritance? First, it is surprising that all life forms use the same alphabet at the genetic level. The barcode of life consists of four letters:

A (for adenine)
T (for thymine)
G (for guanine)
C (for cytosine)

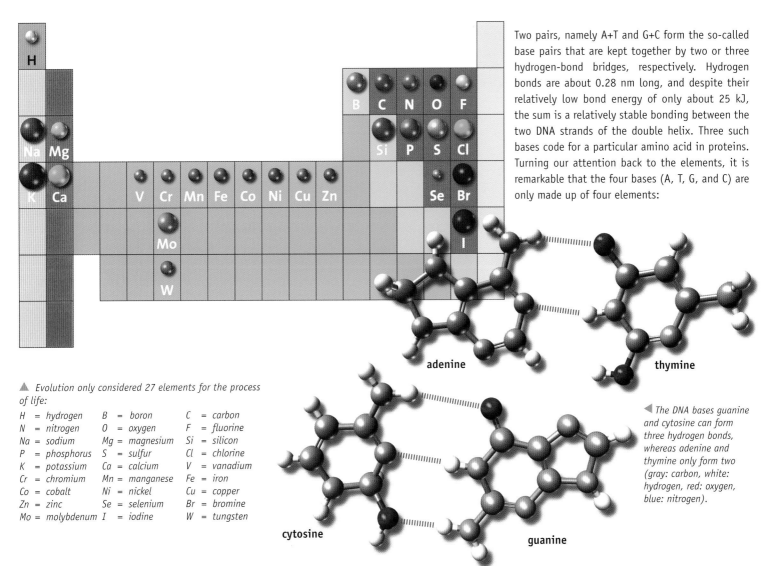

Two pairs, namely A+T and G+C form the so-called base pairs that are kept together by two or three hydrogen-bond bridges, respectively. Hydrogen bonds are about 0.28 nm long, and despite their relatively low bond energy of only about 25 kJ, the sum is a relatively stable bonding between the two DNA strands of the double helix. Three such bases code for a particular amino acid in proteins. Turning our attention back to the elements, it is remarkable that the four bases (A, T, G, and C) are only made up of four elements:

▲ Evolution only considered 27 elements for the process of life:

H = hydrogen	B = boron	C = carbon
N = nitrogen	O = oxygen	F = fluorine
Na = sodium	Mg = magnesium	Si = silicon
P = phosphorus	S = sulfur	Cl = chlorine
K = potassium	Ca = calcium	V = vanadium
Cr = chromium	Mn = manganese	Fe = iron
Co = cobalt	Ni = nickel	Cu = copper
Zn = zinc	Se = selenium	Br = bromine
Mo = molybdenum	I = iodine	W = tungsten

◄ The DNA bases guanine and cytosine can form three hydrogen bonds, whereas adenine and thymine only form two (gray: carbon, white: hydrogen, red: oxygen, blue: nitrogen).

carbon, hydrogen, nitrogen, and oxygen (C, H, N, and O). To form the stable double helix, nature still required phosphoric acid as a connecting link between the special sugar molecules of the molecular strand, hence the element phosphorus was also enlisted.[1]

The mechanistic properties are clearly improved by the incorporation of proteins. Insects and crustaceans have a shell made of chitin, a chain molecule (polymer) based on chemically modified (acetylamino-) glucose.

---

[1] The number 4 recurs remarkably often: carbon, the central element of life, is tetravalent. Life has four criteria. The genetic code is written with four "letters", which in turn are made up of four elements. (The elements themselves occur in four states of matter: solid, liquid, gas, and plasma.) It could, of course, just be a coincidental frequency, as the number 5 is also conspicuously often encountered in life. We have five sense, five fingers and toes. Starfish have five arms. Apples have five carpels. We encounter the number 6 in honeycombs and in numerous examples in the world of crystals, the most prominent being snowflakes. We can only take note of these facts, and even partly explain them. But we must be careful not to draw any deeper conclusions. Or did someone have an idea to combine arithmetic and life?

The building blocks of proteins are the alpha-amino acids, and exclusively those with the L-configuration. There are 20 that occur in nature. They too all consist of the four elements C, H, N, and O; two amino acids additionally contain sulfur (cysteine and methionine). In certain, but vital, enzymes (the peroxidases), sulfur is replaced by selenium.

The formation of hard skeletal structures that give some life forms their shape is a consequence of calcium. Simply said, the shells of lower organisms are generally made up of brittle calcium carbonate and the interior skeletons of higher animals are made up of tough calcium phosphate.

▲ *Insects, such as this impressive longicorn beetle as well as crustaceans form their hard outer shells from chitin, and organic substance.*

Sodium and potassium are used for the electrochemical transfer of signals in the nervous system. The contraction and relaxation of muscles are regulated by an interplay of calcium and magnesium ions. The counterion is generally the chloride ion.

What is the purpose of the many trace elements? Nature placed tight constraints on life. It takes place in a temperature range between 0 °C and about 40 °C (although there are exceptions of up to 100 °C). The pressure is constant at about 1 atmosphere, and the pH value deviates very little from neutral (about pH 7). Nevertheless, an incredible variety of molecules are formed in cells. The secret lies in the fact that during the process of evolution an unimaginable number of biological catalysts came into being; these enzymes allow the production of all these substances. The entire genome of an organism is essentially a huge library of production codes for these catalysts. As enzymes are proteins, they are also only made up of the five elements C, H, N, O, and S. Special tasks require the additional assistance of the complex-forming properties of metals, for example, iron in hemoglobin for oxygen transport, magnesium in chlorophyll as part of the photosynthetic apparatus, cobalt in vitamin B12 for methylation reactions, and molybdenum for nitrogen fixation in the root nodules of legumes such as beans and peas.

There are some exceptions. The diatoms are unicellular microorganisms that protect themselves with a filigree skeleton of silicon dioxide (silicic acid). The sometimes major fossil deposits of kieselguhr (diatomaceous earth or diatomite) are quite well-known. Silicic acid also plays a role in higher life forms. This is also true for boron, which previously had only been known to be important for plants. But it also seems to have a function in animals. Nickel is present in a series of anaerobic microorganisms, but its presence in higher life forms is not certain. Whether antimony belongs to the trace elements is yet to be determined. Mercury is known to be methylated in organisms, but seems to be of no physiological importance. The total number of elements that are known for certain to be involved in metabolic pathways amounts to 27, that is, about a third of the elements in consideration (excluding noble gases and radioactive elements).

Why evolution only made use of these elements cannot be answered. The fact that the Periodic Table holds the possibility of allowing something as wonderful as life to come into existence is and remains a mystery. But not only that: at the end of the development, a being appears on the screen that is able to discover these very elements and to find out about their substructures. And the fact that we can ponder over this is even more mysterious.

▶ *Diatoms (unicellular organisms in water) build their highly symmetrical skeletons out of silica.*

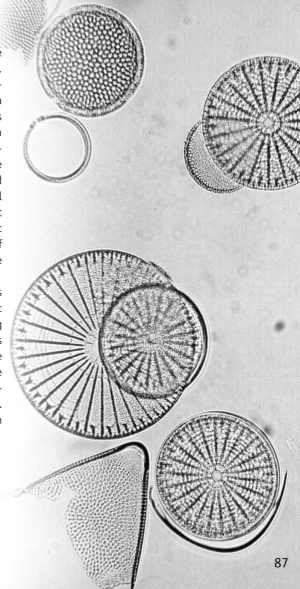

The Elements in Culture and Language

As some elements and their properties were known early on, they often crop up in the cultural heritage of people, that is, in myths, fairytales, and idioms.

Owing to its shine, value, and durability **gold** is the undisputed champion. Everyone knows the saying: "All that glisters is not gold". Before that was "A trade in hand finds gold in every land", which still holds today. Every carefully collected assembly of idioms or quotations would easily include 50 or more examples. It is surprising exactly how often the word "gold" appears in the Book of Books. It occurs 363 times in the bible. And in a further 126 places a derived form (golden or gold coin) is found. The Holy Bible is itself a golden treasure. It is astonishing that the word "hell" is not mentioned once.

As we are in religious territory, let us think of the seven deadly sins defined by the catholic church: pride, envy, greed, lust, gluttony, wrath, and sloth. After these strict standards, some elements do not come off lightly. Gold and the other precious metals are usually seen with a bit of pride. The halogens fall as a result of their greed: they remorselessly pluck electrons from the robes of other elements. The noble gases cannot be outdone in their sloth. But before we go round in circles and start looking for gold in myths, fables, stories, novels, or film titles let us remember the venerable saying: "Speech is silver, but silence is gold".

This takes us to **silver**. Although the word is not rare in pearls of speech, it does crop up much less frequently than gold. The world silver appears 230 times in the bible, and another 101 times in derived words. As the example above shows, silver is often used in a rather more negative context. Let us think of Judas and his 30 pieces of silver. But this beautiful metal does not deserve this reputation. Our modern age has silver to thank for photography and film. Even though this era is coming to an end, the pictures will stay forever. Whether this is true for modern data-storage devices will only be known 100 years from now.

To keep the order, next in line would be the bronze medal. This is known to be an alloy of copper and tin and is thus not really part of this discussion. So let us stick to **copper**. With 114 hits in the bible, this element does, indeed, take third place, but much lower down the rank. That is also true today. Gold and silver wedding anniversaries are celebrated, but copper not.

Iron does come up in idioms, but more seldom. One should "strike while the iron is hot", but one should also "keep many irons in the fire". Everyday pearls of wisdom we all know. But more commonly, reference is made to a special property of iron: its hardness. Examples that speak for themselves include an iron fist, iron will, the Iron Chancellor, and the Iron Lady. When a married couple have remained together ironclad for six years, they celebrate their iron wedding anniversary.

Lead owes its special role to its proverbial heaviness. Its specific weight is similar to that of gold, which is why it was a favorite starting point for alchemists in their quest to produce gold. The Romans considered lead as a symbol for the dark forces of the underworld. Many lead tablets have been found onto which curses have been engraved, something like a Latin voodoo cult. On the other hand, lead had widespread use in the everyday life of the ancient Romans. In particular, wine was fortified in large volumes in lead vessels. This has given rise to the hypothesis that the weakening of the Roman Empire was brought about by insidious lead poisoning. But proof for this suspicion never made many strides, almost as if it had "lead shoes".

Georg Christoph Lichtenberg has suggested that lead had the greatest influence of all the elements on our history. It was excessively abused in the form of lead shot but intensively used in type. The bullet will stay in use at least until more horrific weapons become freely available. In the area of information technology, lead has practically disappeared from the scene. Instead, **silicon** has taken the leading role in microchips. In the printing industry, printers' ink has slowly been edged out by **carbon**, which has pushed its way onto the stage in the form of plastics for coated print rolls. This closes a historical cultural cycle, as the first cave paintings by our forefathers were produced with charcoal. Chemists must wonder why all the elements used by mankind for his information needs are from group 4 (new system: group 14) of the Periodic Table. To explain, or even to justify, this line of thought will be a tough nut for science to crack.

▶ *Although gold has since antiquity been a symbol of power and wealth, it is only now showing its true value in that its resistance to corrosion makes it an ideal material in electronics.*

Chemical Olympics

When Pierre de Coubertin revived the ancient Olympic Games in 1896, it was under the motto: "Citius, Altius, Fortius". That is to say: faster, higher, further (actually: stronger). At that turn of the century, the Periodic Table of the elements was almost complete. One could even have transferred this Olympic concept to the elements. Which atom is the biggest? (Cesium) Which atom takes poll position for ionization energy? (Helium) Which metal has the highest melting point? (Tungsten) There could be any number of questions. Of course, this comparison to the Olympics has one snag: once a winner, always a winner!

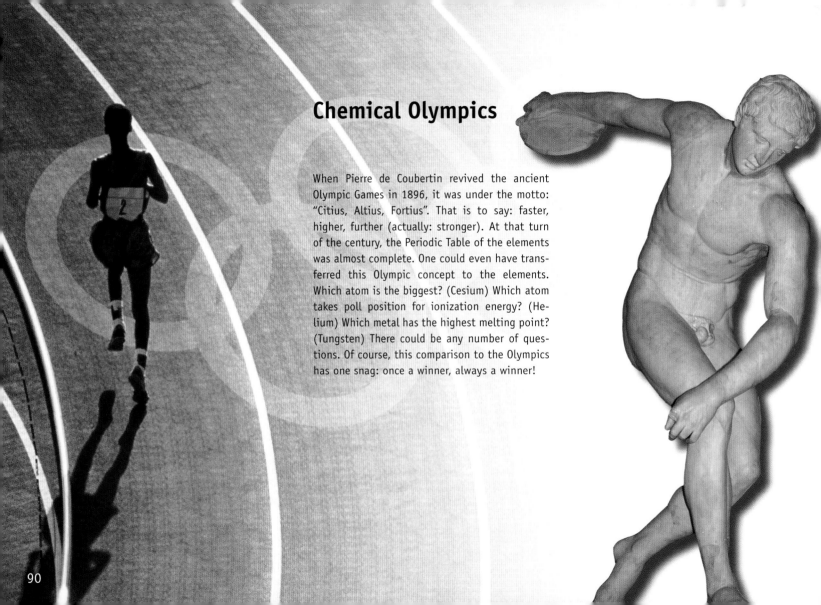

The ten elements already known (but not as elements) in antiquity:

Carbon	charcoal
Gold	jewelry, coins
Silver	jewelry, coins
Copper	tools, weapons, coins
Tin	bronze, coins
Iron	tools, weapons
Lead	anchors, pipes, containers
Mercury	luxury
Antimony	cosmetics
Sulfur	burning

The ten natural elements that were named after places:

Er –	erbium	(*Ytterby*, mining town in Sweden)
Hf –	hafnium	(*Hafnia*, Latin name for Copenhagen)
Ho –	holmium	(*Holmia*, Latin name for Stockholm)
Lu –	lutetium	(*Lutetia*, Latin name for Paris)
Mg –	magnesium	(*Magnesia*, large city in the Middle East)
Re –	rhenium	(*Rhenus*, Latin name for Rhine)
Sr –	strontium	(*Strontian*, town in Scotland)
Tb –	terbium	(*Ytterby*, mining town in Sweden)
Yb –	ytterbium	(*Ytterby*, mining town in Sweden)
Y –	yttrium	(*Ytterby*, mining town in Sweden)

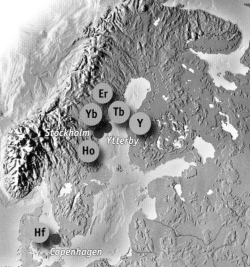

Price:
In the mid-18th century, Napoleon III ate off "precious" aluminum plates, while his guests had to be content with normal gold. The price of the element today is just a question of availability, demand, and speculation! In the case of elements for the semiconductor industry, the purity is an added factor in determining the price.

Malleability:
Gold is the most ductile element. An amount of 1 g of gold can be stretched into a 24-km-long fine wire (actually, a thread). It can also be rolled (or beaten) into 100-nm-thick gold leaf. It then looks blue green in color, and red when light is shone through it.

Heat conductivity:
Diamond is the best heat conductor (four times better than copper). When kissing a lady's hand, if one discretely touches the stone in the ring with one's lips and the diamond feels cold, then it genuine! But don't try to bite it off, as diamond is the hardest substance in the world.

Strength-to-weight ratio:
A titanium wire can hold its own weight up to a length (height) of 25 km before the suspension causes it to rip.

Toxicity:
Beryllium is not the most aggressive of the elements (that is known to be fluorine), but it is considered the most toxic.

Magnetism:
Only three elements are naturally magnetic: iron, cobalt, and nickel.

Isotopes:
Tin hold the record with 10 stable isotopes. There are 19 so-called "pure elements" of which there is only one isotope. These anisotopic elements are: beryllium, fluorine, sodium, aluminum, phosphorus, scandium, manganese, cobalt, arsenic, yttrium, niobium, rhodium, iodine, cesium, praseodymium, terbium, holmium, thulium, gold, and bismuth.

Atomic volume:
Cesium is the most voluminous element and also the most reactive metal.

Relationship:
There are some elements that are so similar to each other, that even the most-practiced experts have problems separating and distinguishing them. They are always found together in their ores, in which they occupy the crystal lattices interchangeably. Conspicuous examples of such chemical siblings include tungsten and molybdenum, niobium and tantalum, as well as zirconium and hafnium.

Suggestion:
Why not visit a mineral collection? Their wealth and diversity makes them the flowers of the mineral kingdom.

The ten elements with the highest melting point:

		Atomic number	
1.	Carbon (diamond)	(6)	3550
2.	Tungsten	(74)	3410
3.	Rhenium	(75)	3180
4.	Osmium	(76)	3045
5.	Tantalum	(73)	2996
6.	Molybdenum	(42)	2617
7.	Niobium	(41)	2468
8.	Iridium	(77)	2410
9.	Ruthenium	(44)	2310
10.	Boron	(5)	2180

The ten elements with the highest boiling point:

		Atomic number	
1.	Rhenium	(75)	5597
2.	Tungsten	(74)	5555
3.	Tantalum	(73)	5420
4.	Hafnium	(72)	5400
5.	Technetium	(43)	5030
6.	Osmium	(76)	5027
7.	Carbon (sublimes)	(6)	4827
8.	Thorium	(90)	4787
9.	Niobium	(41)	4742
10.	Molybdenum	(42)	4639

Certificate
for a gold bar

Der Buyer!

Congratulations on your wise purchase! You have chosen the safest way to invest your savings. Diamonds can burn away as they are made of carbon. Precious stones discolor at high temperatures. Your gold is forever! However, gold exhibits some properties that for legal reasons we bring to your attention. As a reputable firm, we find it our duty to disclose all facts, both positive as well as misunderstandings.

1.
Like all elements, 99.999 999 999 99% of gold is empty space. (Ernest Rutherford, Nobel Prize for chemistry, 1908, can be blamed for this discovery.)

2.
Gold atoms contain 78 protons. Science cannot be sure if theses are infinitely stable. Should a proton decompose into a neutron and an electron, a platinum atom will result, implying a further increase in the value of your investment.

3.
We guarantee a purity of 99.99%. The remainder could be made up of any of the other elements of the Periodic Table. We do not accept responsibility for any consequences.

4.
Gold contains free-moving electrons, which can also reside on the exterior of the bar (by tunneling). We cannot accept liability for any possible consequences.

5.
Gold is quite heavy and exhibits correspondingly strong gravitational forces. It could exercise considerable attraction on undesired elements. Store it safely.

6.
You could increase the weight of your bar considerably by taking it to Jupiter or Saturn. Avoid the moon, where it would only have about one sixth of its current weight.

7.
We cannot provide an opinion as to whether gold consists of strings or superstrings, in which case the gold bar could be considered as a compact batch of energy.

8.
Avoid all contact with antimatter. This could mean the end of your gold and most probably you too!

9.
Keep gold away from aqua regia. Dissolved gold is not quite as valuable.

10.
According to applied Relativity Theory, you own the gold, but the gold has also taken possession of you. Make a wise decision.

Definitions

A mole of a substance is the number of elementary particles (atoms, molecules) found in the mass (in grams) of that substance that corresponds to its atomic or molecular weight. In molar volumes, which amount to 22.4 L for gases and are different from solid to solid, there are always the same number of atoms or molecules. This "magic number", Avogadro's number, is 6.022×10^{23} mol^{-1}.

For example, gold has an atomic weight of 197 and a density of 19.32 g cm^{-3}. This means that 197 g of gold (a cube with 10.2 cm^3 and edges of 2.17 cm) contains 6.022×10^{23} gold atoms. This holds for all elements: a mass in grams that corresponds to its atomic weight contains Avogadro's number of atoms.

Radium

The laboratory notes of Marie and Pierre Curie detailing their discovery of radium still emit such strong radiation that they have to be stored behind lead shields.

Spectral analysis

In the 1859 the chemist Robert Wilhelm Bunsen and his younger colleague, the physicist Gustav Kirchhoff, discovered a surprising phenomenon of spectroscopy. The emission and absorption spectra of an element are identical. They thus put into place an ideal tool for the discovery and identification of elements. Indeed, they themselves discovered cesium (1860) and rubidium (1861). In total, at least 20 elements were found by using spectroscopic techniques (including X-ray spectroscopy).

Masses
Proton: 1.673×10^{-24} g
Neutron: 1.675×10^{-24} g
Electron: 9.109×10^{-28} g

◀ **A proton is 1837 times heavier than an electron.**

Artificial elements

Rhenium (75) was discovered in 1925 by Ida Tacke and Walter Noddack as the last naturally occurring element. The first artificially produced element was identified by Emilio G. Segrè in 1937. Ernest Lawrence detected technetium in a molybdenum sample, which he had bombarded in his cyclotron. All elements discovered since then have been generated artificially.

Fall of a Winged Word

▲ *Leukipp (5th century BC)*

Since the pre-Socrates philosophers, who two and half thousand years ago tried to understand the world with reason, there has been a winged word, that was especially known in its Latin form: Natura non saltat — nature doesn't jump. All experience has shown that in nature one thing always leads to another. Changes occur smoothly and "flow" into each other. Of course, occasionally there is some turbulence, but mostly at the edges and without any influence on the mainstream. The law of causality was also considered to be unyielding in nature: small cause — small effect, large cause — large effect. This concept was firmly engrained in the classical picture of the world. This was true for physics, chemistry, and, above all, biology. Charles Darwin penned the slow but continually adapting changes of his theory of evolution. Even then critics argued that the theory could explain why there were brown bears as well as polar bears, but could not explain why there were bears at all. This question was only solved with the advent of developmental genetics.

But it was Max Planck who shattered the paradigm of the steadiness of nature. He showed that atoms could not absorb energy in all forms and quantities, but only in so-called quanta, that is, in defined amounts. Thus, electrons jump, as we explain it today, from one energy level to another. Natura saltat! Albert Einstein's theory was even more groundbreaking: space and time form a continuum, matter and energy, in contrast, are quantized, essentially "grainy", so to speak. In this case, nature cannot but jump.

Actually, this contradiction of ancient wisdom should have been noticed much earlier by the natural scientists in the 19th century. In fact, it was the chemists who were the closest to bringing down the winged word. When they were in a position to determine atomic weights, the neighborly proximity of many elements was clear to them. What small differences between the atomic weights of carbon, nitrogen, and oxygen and what dramatic differences in their properties. There were no smooth transitions, but really huge jumps. It should all have become crystal clear when the neighbor of fluorine, neon, was discovered. Let us consider this picture: here is the aggressive fluorine that ravenously rips out an electron from the shell of just about any element to make itself comfortable as the fluoride anion. But add a proton and electron (as well as a neutron, as we now know), and the new element is so inert that even today no chemical compound of neon is known. A more obvious jump cannot be imagined. It is a pity that our chemical forefathers did not consider this aspect. After all, it was they who in their research of the Periodic Table brought the time-honored winged word down from its intellectual pedestal, even if they were not aware of this.

NATURA NON SALTAT

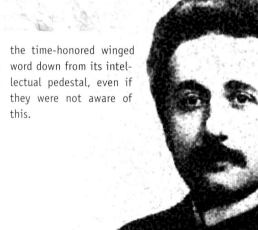

Water

Water — not an element, but elementary for life. For this reason, when astronomers suspect life on distant worlds, they search for water. In this respect, the Earth is paradise. 71% of the surface is covered with water with a volume of 1.386 million km³. This equates to a cube with an edge length of 1110 km, which is almost equivalent to one third of the diameter of the moon. This amount would be plentiful if life on land were not dependent on fresh water. Sea water, with approximately 35g salt/liter, accounts for 96.5% of the total water. In addition, 1% of the total water is salty ground water; this leaves only 2.5% as fresh water. Approximately 69% of this fresh water is trapped as ice and glaciers in the poles. A further 30% is found as ground water and permafrost. Only 1.2% of the surface water (that is 0.3‰ of the total water) is accessible as fresh water. In conclusion, drinking water is globally a limited resource, and for the thirsty, more valuable than gold.

1 gram gold

Gold from sea water – a dream

*Approximately 70 elements can be found in sea water, however, only in minuscule amounts. Among these elements is gold. Unfortunately, the average gold content is around $0.01\,mg/m^3$. There have been, and still are, attempts to isolates this gold, however, the costs are too high. The following example illustrates how difficult it is for us to imagine atomic dimensions. Hypothetically, if 1 g of gold (a cube with an edge length of 0.4 cm) was distributed evenly in the ocean water, then every liter, regardless of where it was collected, would contain at least **two atoms** of this gold.*

Distribution of Global Water

- Oceans 96.5%
- Groundwater (Saline) 1%
- Fresh Water 2.5%
  - Groundwater 30.1%
  - Polar Ice 68.7%
  - Surface water 1.2%

Total water: 1.386 million km³
1 km³ = 1 billion t water

Total fresh water: 34.7 million km³

Interesting Facts about the Periodic Table

▲ Relative atomic sizes (averages)

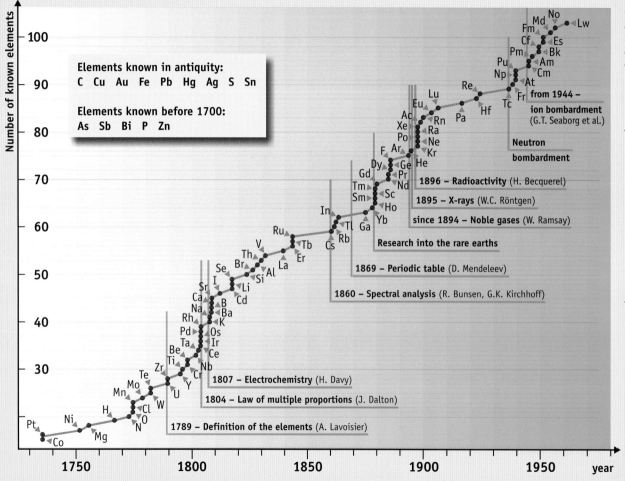

Chronology of the Discovery of the Elements

Elements known in antiquity:
C Cu Au Fe Pb Hg Ag S Sn

Elements known before 1700:
As Sb Bi P Zn

1807 – Electrochemistry (H. Davy)
1804 – Law of multiple proportions (J. Dalton)
1789 – Definition of the elements (A. Lavoisier)
1860 – Spectral analysis (R. Bunsen, G.K. Kirchhoff)
1869 – Periodic table (D. Mendeleev)
Research into the rare earths
since 1894 – Noble gases (W. Ramsay)
1895 – X-rays (W.C. Röntgen)
1896 – Radioactivity (H. Becquerel)
Neutron bombardment
from 1944 – ion bombardment (G.T. Seaborg et al.)

◀ The chronology of the discovery of the elements is a textbook example of scientific development. Three phenomena are particularly clear.

1. Development is laborious and does not occur linearly or predictably.

2. Jumps in knowledge arise from new theoretical concepts, such as the disproving of the phlogiston hypothesis ("the" paradigm shift) or the atom model of Dalton and the Periodic Table. An equally accelerating effect results from the discovery of new methods, such as electrochemistry, spectral analysis, and X-rays.

3. Advances in theory and developments in methodology arose mostly from fundamental research. This aspect was decisive for the research output in Europe and should remain so.

Places of Discovery

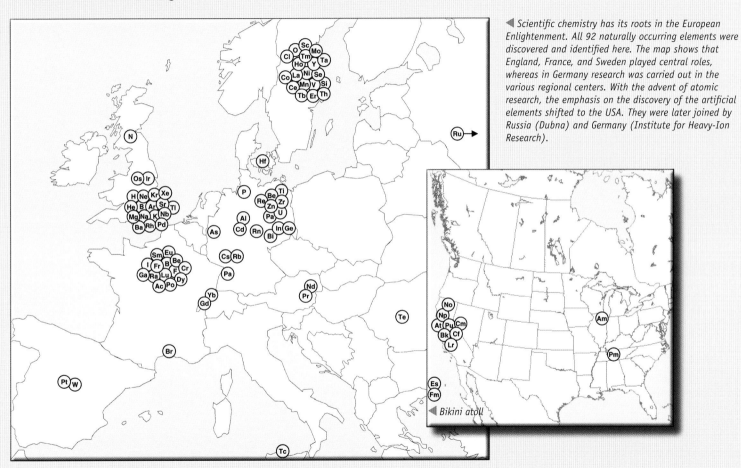

◀ Scientific chemistry has its roots in the European Enlightenment. All 92 naturally occurring elements were discovered and identified here. The map shows that England, France, and Sweden played central roles, whereas in Germany research was carried out in the various regional centers. With the advent of atomic research, the emphasis on the discovery of the artificial elements shifted to the USA. They were later joined by Russia (Dubna) and Germany (Institute for Heavy-Ion Research).

◀ Bikini atoll

Periodicity of Atomic Properties

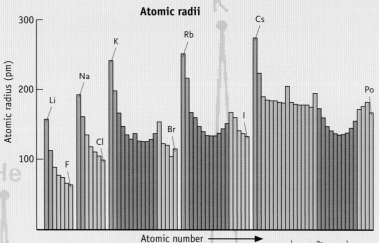

Atomic radii

▲ Atomic radii generally decrease across a period owing to the increase in effective nuclear charge. The radii increase down a group as more outer shells are occupied. The alkali metals have the largest atomic radius of the respective period because there is only one electron in the outer shell.

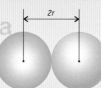

Atomic radius

▲ Johannes van der Waals (1837–1923) defined the atomic radius as half the distance between two atomic nuclei. In reality, the electron clouds have no clear boundaries.

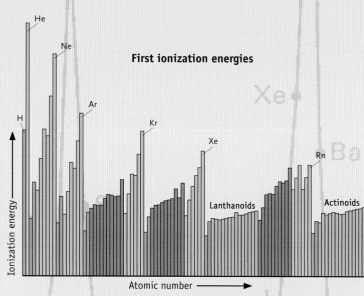

First ionization energies

▲ The noble gases exhibit the highest ionization energies because, according to the octet rule, they have optimal electron configurations. The ionization energies of the alkali metals are correspondingly low.

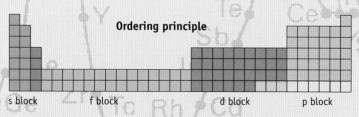

Ordering principle

s block   f block   d block   p block

▲ The Periodic Table can also be ordered according to the electrons in the outer orbital. This makes the "inner occupation" in the case of the transition metals and the lanthanoids and actinoids particularly clear.

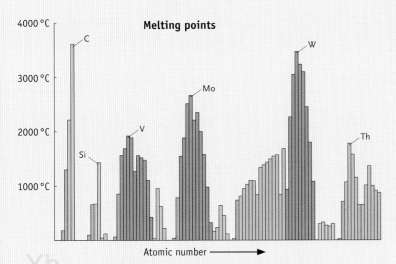

▲ The phenomenon of periodicity is particularly clear in the melting points of the elements. It is however remarkable, because this is a purely physical property. The melting point is not an atomic property, but is determined by the relationships in the crystal lattice. Therefore the maxima and minima do not coincide with the beginning or end of a period as is the case with the atomic radii and ionization energies.

The graph shows that there are no more solid substances in the universe at temperatures above 3600 °C — only smelts, gases, and plasmas.

Comparison of atomic with ionic radius

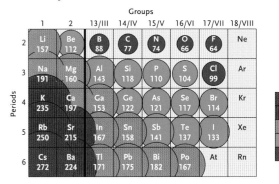

▲ The atomic radii (averages) of the main-group elements (in pm = 10^{-15} m). The radii decrease from left to right across a period and increase down a group. The colors in this picture and the one below represent approximate values according to the scales on the right.

▼ The ionic radii of ionized main-group elements (in pm). Cations have smaller radii and anions larger radii than their corresponding atoms.

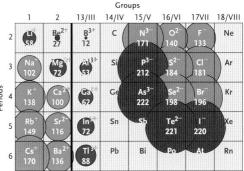
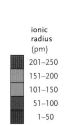

Abundance of the Elements (percentage atoms)

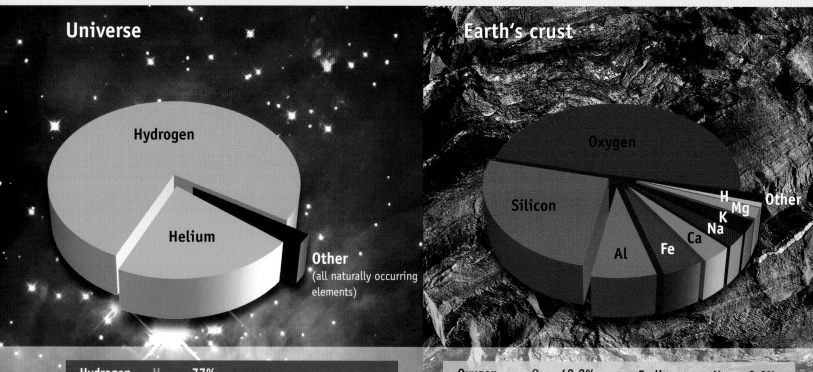

Atmosphere

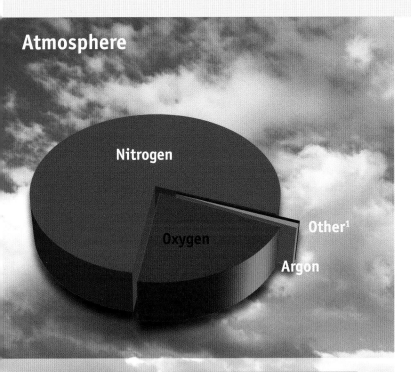

Nitrogen	N_2	78.1%	Argon	Ar	0.9%
Oxygen	O_2	20.9%	Other[1]		0.1%

1 Including carbon dioxide (CO_2) currently at 0.038 %. This trace gas is important for the biosphere and therefore has an important influence on world climate.

Humans

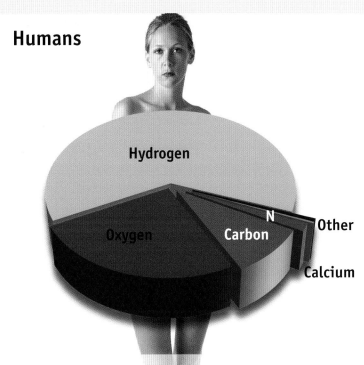

Hydrogen	H	62.9%	Nitrogen	N	1.4%
Oxygen	O	25.5%	Calcium	Ca	0.3%
Carbon	C	9.5%	Other		0.4%

World Production of Elements and Raw Materials (2003)

By economic value (in millions of euro)

#	Material	Value
1.	Natural oil	731 824
2.	Natural gas	289 754
3.	Coal	162 475
4.	Gold	26 625
5.	Iron	17 687
6.	Copper	16 250
7.	Diamond	9 162
8.	Nickel	5 072
9.	Rock salt	4 552
10.	Chromite	4 377
11.	Potash	4 370
12.	Platinum	3 472
13.	Zinc	2 918
14.	Boron minerals	2 805
15.	Bauxite	2 425
16.	Phosphate	2 249
17.	Silver	2 065
18.	Sulfur	1 510
19.	Manganese	1 430
20.	Molybdenum	1 114
21.	Lead	1 083
22.	Cobalt	962
23.	Tin	957
24.	Magnesium	848
25.	Uranium	801
26.	Fluorspar	547
27.	Niobium + tantalum	432
28.	Titanium	401
29.	Zirconium	332
30.	Graphite	321
31.	Vanadium	221
32.	Tungsten	220
33.	Baryte	192

By amount (in 1000 tons, natural gas in millions of m³)

#	Material	Amount
1.	Coal	4 332 671
2.	Natural oil	3 549 100
3.	Natural gas	2 697 900
4.	Iron	640 822
5.	Rock salt	257 148
6.	Bauxite	154 480
7.	Sulfure	68 023
8.	Phosphate	43 496
9.	Potash	28 283
10.	Chromite	19 265
11.	Copper	13 759
12.	Zinc	9 489
13.	Manganese	8 104
14.	Baryte	6 898
15.	Fluorspar	5 370
16.	Boron minerals	4 811
17.	Lead	3 456
18.	Titanium	2 332
19.	Nickel	1 389
20.	Graphite	958
21.	Zirconium	942
22.	Magnesium	499
23.	Tin	256
24.	Molybdenum	133
25.	Tungsten	56
26.	Vanadium	51
27.	Cobalt	46
28.	Uranium	35
29.	Niobium + tantalum	34
30.	Silver	20
31.	Gold	2.64
32.	Platinum metal	0.28
33.	Diamond	0.03

- Precious metals and stones
- Energy raw material
- Industrial minerals
- Metals

▲ This table show the extreme importance of raw materials that serve as sources of energy. In the meantime, the demand and prices have increased further. In 2005, there was a demand for 3.92 billion tons of oil. At an average price of 55 US $ per barrel, this corresponds to 1200 billion US $ (1000 billion €).

▲ The demand for energy raw materials is several times greater (also in terms of amount) than that for the rest of the elements. Only uranium is additionally used for energy production. The ratio of the three energy raw materials (brown coal was not calculated) to the 35 000 tons of uranium supplied gives 300 000:1.

◀ To visualize the ratio of the amounts, the total demand for uranium is represented as a red sphere (left; of that, only 0.71 % is fissionable $^{235}U = $ ⬤). The corresponding amount of the energy raw materials is represented by the black sphere in the background, of which only a portion is shown.

Do we have enough resources?

An astronaut from another planet would in particular notice one phenomenon on Earth. The staggering increase in the population. In the year 2006 we passed the 6.6 billion-people mark. The 9 billion figure is inevitable. After 2050, there could be 10 billion people. The more people live in comfortable environments, the greater the demand for resources will be. One element is at the center of this discussion: carbon.

Carbon plays a double role. It is the most important energy source (93 % of consumed materials) as well as starting material in chemistry, from which many consumer products are manufactured (but only require 7 % of the natural oil consumed).

The sources for carbon are coal, natural oil, and natural gas. Their consumption has two important problematic consequences. First, the reserves are finite; this means that sufficient alternatives must be found. Second, the combustion of these resources releases CO_2. It is not certain how much the atmosphere can take before serious or irreparable damage is caused. With the other elements there are a few bottlenecks, which, however, can be circumvented by recycling and by fixing the price. The question of carbon, on the other hand, is of a fundamental dimension. Carbon has become not only the central element of life, but development has also made it a key factor for civilization.

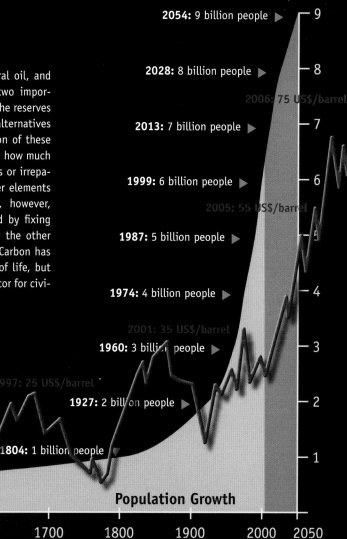

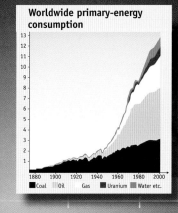

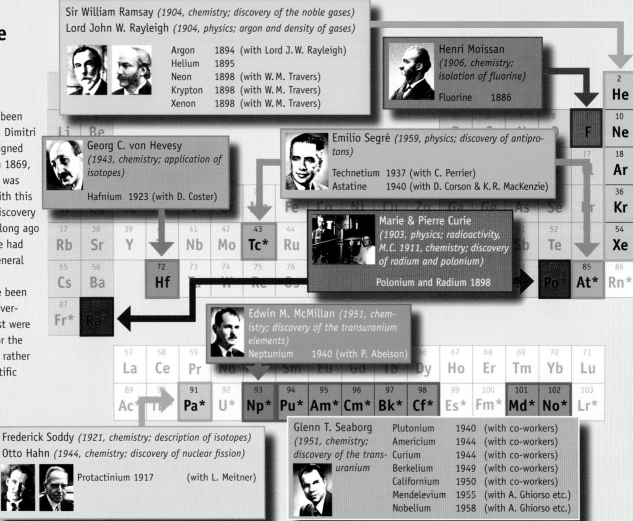

Nobel Prize Winners

The Nobel Prize has been awarded since 1901. Dimitri Mendeleev, who designed the Periodic Table in 1869, lived until 1907. He was not distinguished with this prize, because his discovery had been made too long ago and in the meantime had become scientific general knowledge.

Nobel Prizes have been awarded to 11 discoverers of elements. Most were not distinguished for the discovery itself, but rather for a different scientific achievement.

Sir William Ramsay *(1904, chemistry; discovery of the noble gases)*
Lord John W. Rayleigh *(1904, physics; argon and density of gases)*

Argon	1894	(with Lord J.W. Rayleigh)
Helium	1895	
Neon	1898	(with W. M. Travers)
Krypton	1898	(with W. M. Travers)
Xenon	1898	(with W. M. Travers)

Henri Moissan *(1906, chemistry; isolation of fluorine)*
Fluorine 1886

Georg C. von Hevesy *(1943, chemistry; application of isotopes)*
Hafnium 1923 (with D. Coster)

Emilio Segrè *(1959, physics; discovery of antiprotons)*
Technetium 1937 (with C. Perrier)
Astatine 1940 (with D. Corson & K. R. MacKenzie)

Marie & Pierre Curie *(1903, physics; radioactivity, M.C. 1911, chemistry; discovery of radium and polonium)*
Polonium and Radium 1898

Edwin M. McMillan *(1951, chemistry; discovery of the transuranium elements)*
Neptunium 1940 (with P. Abelson)

Frederick Soddy *(1921, chemistry; description of isotopes)*
Otto Hahn *(1944, chemistry; discovery of nuclear fission)*
Protactinium 1917 (with L. Meitner)

Glenn T. Seaborg *(1951, chemistry; discovery of the transuranium)*

Plutonium	1940	(with co-workers)
Americium	1944	(with co-workers)
Curium	1944	(with co-workers)
Berkelium	1949	(with co-workers)
Californium	1950	(with co-workers)
Mendelevium	1955	(with A. Ghiorso etc.)
Nobelium	1958	(with A. Ghiorso etc.)

Alternative Representations of the Periodic System

What is the best didactic and graphic way of representing the periodic system? The representation dating back to Dmitri Mendeleev has successfully stood the test of time. Nevertheless there have always been and still are attempts to represent the relationships in different formats. This task always poses one particular challenge: The occupation of the orbitals by electrons should be clearly recognizable. In most cases, however, the overview is lost. Some proposals are represented here. Some are quite new, which shows that the search has not yet ended.

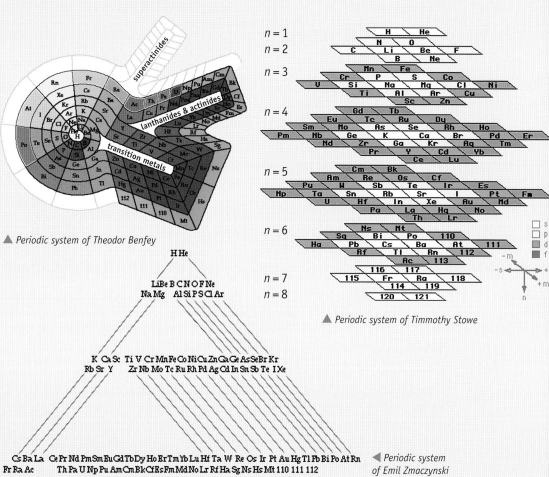

▲ Periodic system of Theodor Benfey

▲ Periodic system of Timmothy Stowe

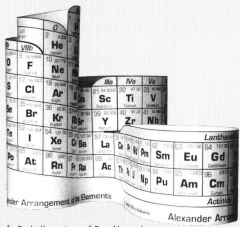

▲ Periodic system of Roy Alexander

◀ Periodic system of Emil Zmaczynski

The Occupation of the Shells and Orbitals of the Elements of the Periodic Table

The beautiful Bohr atomic model is, unfortunately, too simple. The electrons do not follow predetermined orbits. Only population probabilities can be given, which are categorized as shells and orbitals. The orbitals can only accommodate two electrons. Shells and orbitals can also merge ("hybridization"). In the case of carbon, the 2s orbital and the three 2p orbitals adopt a configuration in the shape of a tetrahedron. Each of these sp³ orbitals is occupied by one electron. This gives rise to the sterically directed four-bonding ability of carbon.

The octet rule states that atoms strive to have eight electrons on the outer shell, as is the case with the stable noble gases. One way of doing this is by giving up or taking in electrons. As this gives rise to charged ions, the type of bonding is called ionic. The other possibility is that two atoms share the electron pairs. This means that each then has a quasi-noble-gas configuration. This tendency is particularly pronounced in carbon, for which covalent bonding is predominant in its chemistry.

The shells play a leading role in the structure of the Periodic Table. This graphic representation is borrowed from the Bohr atomic model. Historically, the shells were assigned letters, nowadays they are represented with numbers: shell 1: K; shell 2: L; shell 3: M; shell 4: N; shell 5: O; shell 6: P; shell 7: Q. The orbitals are still conventionally labeled with letters: s, p, d, and f.

The shells have an increasing number of orbitals and can accommodate a correspondingly increasing number of electrons. The first shell (K) accommodates 2, the second 8, the third 18, and the fourth shell up to 32 electrons.

As the atomic number increases, so does the positive charge of the nucleus, and the electrons are bound with a higher energy. However, this increase is not linear. For example, the electrons in the d orbital of the third shell have a higher energy than those in the s orbital of the fourth shell, and hence the latter are filled first. The consequence is the unexpected behavior of the first ten transition elements. In the case of the actinides and lanthanides, even more inner orbitals are occupied. Nature is not so simple, but the scheme should help to visualize this complex structure. And if one can assign the electrons of an element, one is a step closer to successfully unraveling the mysteries of the Periodic Table.

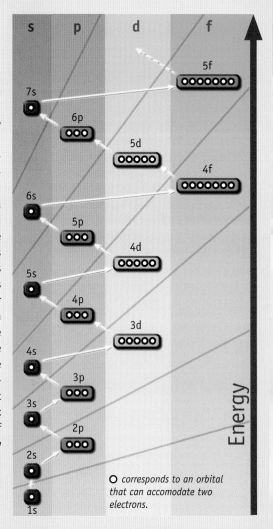

O *corresponds to an orbital that can accomodate two electrons.*

The World of the Elements — Literature

General Literature on the Periodic Table

Atkins, Peter W.;
The Periodic Kingdom:
A Journal of the
Chemical Elements.
Basic Books, New York,
USA (1997)

Ball, Philip;
The Elements: A Very
Short Introduction.
Oxford University Press,
New York, USA (2004)

Ball, Philip;
The Ingredients:
A Guided Tour of the
Elements.
Oxford University Press,
New York, USA (2003)

Bowden, Mary Ellen;
Chemical Achievers,
Chemical Heritage
Foundation (1997)

Emsley, John;
The Elements of Murder:
A History of Poison.
Oxford University Press,
New York, USA (2006)

Emsley, John;
Nature's Building
Blocks, An A–Z Guide
to the Elements.
Oxford University Press
(2001)

Enghag, Per;
Encyclopedia of the
Elements. Wiley-VCH,
Weinheim (2004)

Fluck, Ekkehard;
Heumann, Klaus G.;
The Ultimate Periodic
Table (Poster).
Wiley-VCH, Weinheim,
Germany (2007)

Gonick, Larry;
Criddle, Craig;
The Cartoon Guide to
Chemistry.
Harper Resource,
New York (2005)

Heiserman, David L.;
Exploring the Chemical
Elements and Their
Compounds.
McGraw-Hill, New York,
USA (1991)

Holum, John R.;
Elements of General
Organic and Biological
Chemistry, 9th Ed. Wiley,
New York, USA (1995)

Ihde, Aaron J.;
The Development of
Modern Chemistry.
Dover (1984)

Jones, Loretta;
Atkins, Peter W.;
Chemical Principles:
The Quest for Insight,
3rd Ed. W. H. Freeman,
New York, USA (2004)

Marshall, James;
Discovery of the Elements.
Pearson Custom
Publishing (1998)

Quadbeck-Seeger,
Hans-Jürgen (Ed);
World Records in
Chemistry.
Wiley-VCH, Weinheim,
Germany (1999)

Scerri, Eric R.;
Edwards, Jacob;
Bibliography of
Secondary Sources on
the Periodic System of
the Chemical Elements.
Foundation of Chemistry 3, *183–196* (2001)

Stwertka, Albert;
Guide to the Elements.
Oxford University Press
(1998)

Weeks, Mary Elvira;
Discovery of the
Elements. Journal of
Chemical Education,
Easton, PA. (1960)

Wiberg, Nils; Holleman,
A.F.; Wiberg, Egon (Eds);
Inorganic Chemistry.
Academic Press, London
UK (2001)

General Literature on the Periodic System

Sacks, Oliver;
Uncle Tungsten:
Memories of a Chemical
Boyhood.
Random House, New York,
2001

Djerassi, Carl;
Hoffmann, Roald;
Oxygen. Wiley-VCH,
Weinheim (2001)

Levi, Primo;
Das periodische System.
Süddeutsche Zeitung/
Bibliothek, München
(2004)

Internet

An abundance of important and interesting information can be found on the internet when searching under the keywords "Periodic Table". A good starting point is the homepage of the various chemical societies:
American Chemical Society (ACS):
www.acs.org
Gesellschaft Deutscher Chemiker (GDCh; German Chemical Society):
www.gdch.de
Royal Society of Chemistry (RSC):
www.rsc.org

Various firms, universities, research institutes, schools, and private people have made various lecture notes and information material available online. **Go searching and be surprised!**

Sources of Photographs

Pages 2/3:
DirectMedia; Corbis Dig. Stock
Page 7:
Emilio Segre Visual Archives
Pages 10/11:
William Blake, The Ancient of Days, 1794; Real-Encyklopädie, 11th ed. (1864–68), Ed. Friedrich Arnold Brockhaus (1772–1823)
Pages 16/17/22:
NASA
Page 23:
Bayer AG
Pages 26/27:
NASA; Daimler-Chrysler AG; Photodisc
Pages 28/29:
Photodisc; BASF AG
Pages 30/31:
BASF AG; K+S AG; Photodisc
Pages 32/33:
Photodisc; Ingram Publishing; Hemera Technologies; G. Schulz
Pages 34/35:
Getty Images; Hemera Technologies; cc-Vision

Pages 36/37:
Photodisc; K+S AG; BASF AG; Hemera Technologies
Pages 38/39:
BASF AG; UNESCO; Photodisc, Hemera Technologies
Pages 40/41:
BASF AG; Hemera Technologies
Pages 42/43:
Getty Images; Hemera Technologies; Photodisc; cc-Vision
Pages 44/45:
Hemera Technologies; K+D AG; Photodisc
Pages 46/47:
Hemera Technologies; G. Schulz; cc-Vision, Knoll AG
Pages 48/49:
Hemera Technologies; Knoll AG
Pages 50/51:
Corbis Digital Stock; Bosch AG; Thyssen Krupp AG; NASA
Pages 52/53:
Hemera Technologies ; BASF AG
Pages 54/55:
Hemera Technologies; Photodisc; G. Schulz

Pages 56/57:
Getty Images; cc-Vision; Hemera Technologies; BASF AG
Pages 58/59:
Philips AG; Getty Images
Pages 60/61:
Hemera Technologies; Getty Images; NASA
Pages 62/63:
Siemens AG; Hemera Technologies; Osram AG
Pages 64/65:
Photodisc; Corbis Digital Stock; Getty Images
Pages 66/67:
BASF AG; Wikipedia; Hemera Technologies; Digital Vision
Pages 68/69:
Osram AG; Hemera Technologies; BASF AG; Bosch AG
Pages 70/71:
Photodisc; BASF AG; Hemera Technologies, cc-Vision
Pages 72/73:
Getty Images; Hemera Technologies

Pages 74/75:
US Department of Defense; Getty Images; Digital Vision
Pages 78/79:
NASA; Siemens AG
Pages 80/81:
Emilio Segre Visual Archives
Page 83:
Wikipedia
Pages 84/85:
Photodisc
Pages 86/87:
Hemera Technologies; Photodisc
Pages 88/89:
Getty Images; cc-Vision
Pages 90/91:
Getty Images, Corbis Digital Stock; Mountain High Maps
Pages 92/93:
Hemera Technologies; cc-Vision
Pages 96/97:
Hemera Technologies; MEV Verlag; dpunkt; Ingram Publishing

Pages 98/99:
BASF AG
Pages 100/101:
Wiley-VCH
Pages 102/103:
NASA; cc-Vision

The sources of all photographs have been researched and credited with the greatest care. We apologize if we have made any mistakes in this regard and ask that you inform us.

Chemical Elements and Their Properties According to Atomic Number

AN[1]	Name/Symbol		Atomic weight g/mol⁻¹	Density at 20°C (g/cm⁻³)[2]	Melting point (°C)[3]
1	Hydrogen	H	1.008	*0.084*	*−252.9*
2	Helium	He	4.003	*0.17*	*−268.9*
3	Lithium	Li	6.941	0.53	180.5
4	Beryllium	Be	9.012	1.85	1278
5	Boron	B	10.81	2.46	2180
6	Carbon	C	12.01	3.51	3550
7	Nitrogen	N	14.01	*1.17*	*−195.8*
8	Oxygen	O	16.00	*1.33*	*−183.0*
9	Fluorine	F	19.00	*1.58*	*−188.1*
10	Neon	Ne	20.18	*0.84*	*−246.0*
11	Sodium	Na	22.99	0.97	97.8
12	Magnesium	Mg	24.30	1.74	648.8
13	Aluminum	Al	26.98	2.70	660.4
14	Silicon	Si	28.09	2.33	1410
15	Phosphorus	P	30.97	1.82	44.1(w)
16	Sulfur	S	32.07	2.06	119.6
17	Chlorine	Cl	35.45	*2.95*	*−34.6*
18	Argon	Ar	39.95	*1.66*	*−185.7*
19	Potassium	K	39.10	0.86	63.3
20	Calcium	Ca	40.08	1.54	839
21	Scandium	Sc	44.96	2.99	1539
22	Titanium	Ti	47.87	4.51	1660
23	Vanadium	V	50.94	6.09	1890
24	Chromium	Cr	52.00	7.14	1857
25	Manganese	Mn	54.94	7.44	1244
26	Iron	Fe	55.85	7.87	1535
27	Cobalt	Co	58.93	8.89	1495
28	Nickel	Ni	58.69	8.91	1453
29	Copper	Cu	63.55	8.92	1083.4
30	Zinc	Zn	65.41	7.14	419.6
31	Gallium	Ga	69.72	5.91	29.8
32	Germanium	Ge	72.64	5.32	937.4
33	Arsenic	As	74.92	5.72	613 (subl.)
34	Selenium	Se	78.96	4.82	221
35	Bromine	Br	79.90	*3.14*	*58.8*
36	Krypton	Kr	83.80	*3.48*	*−152.3*
37	Rubidium	Rb	85.47	1.53	38.9
38	Strontium	Sr	87.62	2.63	769
39	Yttrium	Y	88.91	4.47	1523
40	Zirconium	Zr	91.22	6.51	1852
41	Niobium	Nb	92.91	8.58	2468
42	Molybdenum	Mo	95.94	10.28	2617
43	Technetium	Tc	98.91	11.49	2172
44	Ruthenium	Ru	101.1	12.45	2310
45	Rhodium	Rh	102.9	12.41	1966
46	Palladium	Pd	106.4	12.02	1554
47	Silver	Ag	107.9	10.49	961.9
48	Cadmium	Cd	112.4	8.64	320.9
49	Indium	In	114.8	7.31	156.6
50	Tin	Sn	118.7	7.29	232.0
51	Antimony	Sb	121.8	6.69	630.7
52	Tellurium	Te	127.6	6.25	449.5
53	Iodine	I	126.9	4.94	113.5
54	Xenon	Xe	131.3	*4.49*	*−107.1*
55	Cesium	Cs	132.9	1.90	28.4
56	Barium	Ba	137.3	3.65	725
57	Lanthanum	La	138.9	6.16	920
58	Cerium	Ce	140.1	6.77	798
59	Praseodymium	Pr	140.9	6.48	931
60	Neodymium	Nd	144.24	7.00	1010
61	Promethium	Pm	146.9	7.22	1080
62	Samarium	Sm	150.4	7.54	1072
63	Europium	Eu	152.0	5.25	822
64	Gadolinium	Gd	157.2	7.89	1311
65	Terbium	Tb	158.9	8.25	1360
66	Dysprosium	Dy	162.5	8.56	1409
67	Holmium	Ho	164.9	8.78	1470
68	Erbium	Er	167.3	9.05	1522
69	Thulium	Tm	168.9	9.32	1545
70	Ytterbium	Yb	173.0	6.97	824
71	Lutetium	Lu	175.0	9.84	1656
72	Hafnium	Hf	178.5	13.31	2227
73	Tantalum	Ta	180.9	16.68	2996
74	Tungsten	W	183.8	19.26	3410
75	Rhenium	Re	186.2	21.03	3180
76	Osmium	Os	190.2	22.61	3045
77	Iridium	Ir	192.2	22.65	2410
78	Platinum	Pt	195.1	21.45	1772
79	Gold	Au	197.0	19.32	1064.4
80	Mercury	Hg	200.6	*13.55*	*356.6*
81	Thallium	Tl	204.4	11.85	303.5
82	Lead	Pb	207.2	11.34	327.5
83	Bismuth	Bi	209.0	9.80	271.3
84	Polonium	Po	209.0	9.20	254
85	Astatine	At	210.0		302
86	Radon	Rn	222.0	*9.23*	*−61.8*
87	Francium	Fr	223.0		27
88	Radium	Ra	226.0	5.50	700
89	Actinium	Ac	227.0	10.07	1050
90	Thorium	Th	232.0	11.72	1750
91	Protactinium	Pa	231.0	15.37	1554
92	Uranium	U	238.0	18.97	1132.3
93	Neptunium	Np	237.0	20.48	640
94	Plutonium	Pu	244.1	19.74	641
95	Americium	Am	243.1	13.67	994
96	Curium	Cm	247.1	13.51	1340
97	Berkelium	Bk	247.1	13.25	986
98	Californium	Cf	251.1	15.1	900
99	Einsteinium	Es	252.1		(860)
100	Fermium	Fm	257.1		
101	Mendelevium	Md	258.1		
102	Nobelium	No	259.1		
103	Lawrencium	Lr	262.1		

[1] Atomic number
[2] Density: g cm⁻³, in the case of gases: g L⁻¹ *(italics)*
[3] In the case of liquid and gaseous elements: boiling point *(italics)*